NORTHAMPTON COLLEGE
R28105P 0098

Classic Plastics

classic plastics

FROM BAKELITE TO HIGH-TECH

with a collector's guide

Sylvia Katz

with 188 illustrations, 111 in colour

Thames and Hudson

Filmset in Great Britain by Foremost Typesetting,
London

Printed and bound in Japan by Dai Nippon

Acknowledgments

My thanks are due to Marie Aflalo, Pierre Boogaerts, Patrick Cook, Marcel Fleiss, John Hayes, BIP, Veronica Manussis, Barbara Morris, Roger Newport, Anna Rabolini, John Ratcliffe, PRI, Michel Roudillon, Ella Stewart, Pat Thornton; and special thanks to John Kaine, Dr C. A. Redfarn, M.A.W. and the staff of Thames and Hudson.

Note
The author would be pleased to hear from anyone with additional or more accurate information on any of the subjects discussed in this book.

Contents

Introduction

Plastics are materials which are softened by heat and set into lasting form when shaped in a mould. Some are natural; some are semi-synthetic, the result of chemical action on a natural substance; some are synthetic, built up from the constituents of oil or coal. All are based on the chemistry of carbon, with its capacity for forming chains. The molecules that compose them (monomers) link together in the setting or curing process to form chains (polymers), which give plastics their flexible strength. Some plastics retain their ability to be softened and reshaped: like wax, they are thermoplastic. Others set permanently in the shapes they are given by heat and pressure: like eggs, they are thermosetting.

This book is a celebration of plastics. From industrial beginnings in the nineteenth century, plastics have struggled through a hundred and twenty years of glory, failure, disrepute and suspicion on the slow road to public acceptance. Now, at last, one can positively say that plastics are appreciated and enjoyed for what they are; that they make modern life richer, more comfortable and convenient, and also more fun. Plastics are warm materials, sympathetic to the human touch, and their transformation into things that come into contact with human beings is entirely appropriate.

The fact that there are plastics antiques comes as a shock to most people. How can a material that seems so essentially twentieth-century, and one that is so much associated with cheap, disposable products, have a history at all? It is a young technology, and a great part of the fun of collecting plastics is that beautiful pieces of historical interest can still be found very cheaply.

Until recently, the only plastics to find their way into the official auction rooms have been shellac union cases, made to protect early photographs, and turn-of-the-century clocks or furniture, usually inlaid with casein. Now, slowly, classics from the production line are creeping in, such as moulded chairs by Panton and Saarinen and even anti-design objects from the 1970s. Designs by Sottsass, Superstudio and Archizoom from the exhibition 'Italy: The New Domestic Landscape', held in New York in 1972, reached the auction house in Monte Carlo in 1983.

Unlike rare and expensive natural materials, plastics are universally available. They are, as Roland Barthes described them in *Mythologies,* 'the first magical substance that consents to be prosaic'. Plastics are truly magical because they are created by pure alchemy. Who would have expected the debris of rotting organic matter thousands of years old, mined from the earth in the form of crude oil, coal and gas, to metamorphose into satellite wings and silver-coated film for space shuttles?

A design in plastics which has been carefully worked out and manufactured possesses an inherent beauty reflecting its own technology, which goes beyond the function and usefulness of the object – as collectors of such venerable trade names as Bandalasta and

Bakelite will know. The final form of a design is the result of a succession of problems and solutions involving many people.

Each moulding bears the scars of its manufacture: flow lines, witness marks, flash and ejection marks. The designer has shrinkage and creep-resistance to consider, and the material must be warp-resistant, nowadays flame-resistant, and must possess appropriate electrical properties. Appearance is maybe only fifty per cent aesthetic. The fact that, until recently, plastics products invariably had rounded corners was not due to the designers' stylistic preferences but to the structural weakness of sharp edges, prone to chip. Newer polymers are becoming so strong that sharp angles are now no problem. Plastics designs taper because they are easier and cheaper to remove from moulds; split moulds which allow for undercuts are expensive. Thickening the walls of a moulding is not necessarily the answer to a strength problem. Curved forms are far stronger than flat shapes, as stresses are distributed; ribs can be used to stiffen a flat area and disguise flow marks, and can also be exploited aesthetically.

Early plastics

For centuries we have had natural plastic materials which can be softened by heat and pressed into shape. Amber, horn, bone, tortoiseshell, shellac, and moulding compounds made from natural ingredients, papier mâché and bois durci, are the earliest natural plastics: they have a similar molecular structure to the later synthetics, with long linked chains based on carbon.

Horn is a natural plastic which came into very early use. It behaves like a typical thermoplastic sheet material when it is heated and split open, softened again in boiling water or alkaline solution, and then flattened and shaped in a hot press. Layers can be laminated together, as celluloid was later, or the sheet can be pressed into wooden moulds for shaping snuff boxes, bonbonnières or buttons, all of which can now be found quite easily. Ground to powder and mixed with a binder such as dried blood, horn can be compression moulded into more solid buttons, brooches and walking-stick handles, often mounted with silver.

Horn becomes transparent and pale yellow when heated and pressed very thin, a property exploited in the eighteenth century when it was used as a cheap substitute for glass. This was a period when glass window panes were so precious that they were handed down as family heirlooms. Could this have been the beginning of the notion of cheap plastics substitutes? In the nineteenth century horn too became rare and was replaced by semi-synthetic plastics such as casein and celluloid. Twentieth-century sets of nesting tumblers in plastics are still known as 'nests of horns' after their natural horn predecessors.

For centuries the chief use of horn was for making combs. Originally, each tooth was cut carefully by hand from flat sheet with a saw, but fortunately an 'intersecting machine' was developed which could cut four combs from one piece of horn. Its cutter was able to cut two lengths with teeth interlocked, and then each length was cut in half again. Now a dozen finished combs can be injection moulded every few seconds in a single operation.

From 1770 to 1880 horn was the most popular material for making some of the most magnificent combs ever created. Towering hairstyles became the fashion and designs grew more and more intricate and outrageous. These tall combs were either cut by hand or punched out by a cutter, and then hand-painted, inlaid and polished. One of the finest collections can be seen in the Musée du Peigne et des Plastiques in Oyonnax, the plastics centre of France, tucked away in the Jura mountains, and well worth a visit. The museum also has a collection of early plastics processing machinery.

Tortoiseshell, another natural plastic, is actually not the shell of a tortoise but that of the hawksbill turtle. It can be cut and pressed into shape like horn, but always in such a way as to maintain the beautiful natural pattern.

Older than any animal substance, amber is a hard natural thermoplastic resin fossilized from the trees of the species *Pinites succinifer* that flourished 40-60 million years ago, mainly along the Baltic coast. The ancient Greeks used amber as a protective coating, and like shellac it was mixed into a variety of lacquers. It is found in all shades of yellow. The Greeks and Romans shaped it into ornaments, and later it was carved into beads and rosaries. It is a good insulator, which explains its use as cigarette holders and pipe mouthpieces, much better than dark, unpleasant-tasting phenolic Bakelite.

The economic advantage of amber, like horn, is that all waste material can be re-cycled by melting together small pieces under pressure. Such a material was known as Ambroid, and pressed or extruded shapes can be distinguished from real amber by the flow lines visible in the resin. Any injection-moulded amber cigarette holders must be post-1904, when Gaylord patented a machine for their production.

Lacquer work was invented by the Chinese around 1000 BC, making use of the resin from their native lacquer trees, and the technique later spread further afield. Like polyurethane today, Chinese lacquer forms an extremely durable, tough, waterproof coating, and even armour was lacquered for protection. Until cheap plastics appeared in the 1950s, it was used to coat domestic tableware.

The resin is tapped from the tree, normally the species *Rhus verniciflua,* and then strained. On exposure to oxygen, polymerization takes place and the lacquer cures and hardens like a synthetic plastic. The lacquer therefore has to be brushed on in thin layers (some pieces have been known to have 200 coats), and when built up to the required thickness it can be carved. The finished product could later be inlaid with gold, silver and mother-of-pearl.

Shellac is made from lacquer obtained from the secretion of the tiny insect *Coccus lacca,* which inhabits tropical acacia trees. Like amber and horn, it is thermoplastic, so that all scrap, such as broken gramophone records and dental plates, can be ground up and re-moulded. It was patented as a moulding material in 1854 by Samuel Peck in America, who mixed it with woodflour (fine sawdust) and made it into glossy black frames and carrying cases, known as union cases, for 'wet plate' photographs, called ambrotypes, and for daguerreotypes. The ability of shellac to reproduce very fine details was exploited by the record industry; the Berliner label was the first to use shellac in the 1880s, followed later by Edison. The 1850s to 1880s was the heyday for union cases, especially those moulded by Samuel Peck and by Littlefield, Parsons & Co. Shellac is still used to stiffen felt and bowler hats, and is added to wax to make sealing wax.

Papier mâché tends to be relegated to the school art room nowadays, but it is a strong, heatproof and mouldable thermosetting material. It was patented in Birmingham in 1772 by Henry Clay. Layers of paper and glue are placed in a metal mould and dried in an oven. The moulded shape can be sanded and even polished, but is most often found lacquered and inlaid. In the mid-nineteenth century it was often used for tea trays, pen trays, work- and writing-boxes, and was inlaid with mother-of-pearl, or gilt and japanned. In the days of the snuff box, papier mâché made the most economical material for fabricating a box with a tight-fitting lid, but these boxes have become rare.

Bois durci is a rather unusual material, reputedly produced only between 1855 and the late 1880s. It was patented in France and is a mixture of woodflour and albumen from eggs or blood. The dry mixture can be heated and pressed into simple shapes which cannot be re-fashioned: bois durci is thermosetting. Plaques are the most common mouldings in this material, and were primarily used to decorate furniture. They usually portrayed notable figures such as Victoria and Albert, Byron or Shakespeare. Other uses included knife handles, dominoes and jewellery, and the original patent described a technique for producing a moulded-in decoration in metal.

Gutta percha is another material most people today have never heard of, yet in the nineteenth century it was moulded into a great many domestic and industrial products. It is a completely natural plastic, stripped in solid form from the bark of the Malayan palaquium tree. First seen in Europe in 1822, by the 1840s it had been used to make everything from garden hoses to acid bottles, and from jewellery to picture-frames and matchboxes. Entire pieces of gutta percha furniture were exhibited at the London Great Exhibition of 1851, but the material lacks durability and very few objects remain. One continuing use is for the shells of golf balls.

Semi-synthetics

In the nineteenth century scientists started to discover that natural substances reacted to various chemicals to produce new, semi-synthetic substances with very useful properties. In 1838 Charles Goodyear discovered that by combining latex or natural rubber with varying amounts of sulphur, he could make the strong and resilient rubber we now use in moulded form, and which he called vulcanized rubber. The more sulphur he used, the harder the product, until he arrived at the hard, shiny material known in Britain as ebonite or vulcanite and in America simply as hard rubber. Naturally dark in colour, it can be shaped into highly decorative pieces which have a particular beauty of their own. It was often used as a substitute for jet, but arguments over this point are best avoided in street markets.

Vulcanite can still be found quite easily in the form of fountain pens, vesta matchboxes, pipe stems, dressing-table boxes, candlestick telephones and jewellery. Originally the major use for this material was dental plates; the moulded surface could hold just enough pink pigment to pass as gums, and in some countries this application survived as late as the 1940s.

The next major semi-synthetic plastic to appear, and now perhaps the most sought-after by collectors, was Parkesine. In the 1840s the English inventor Alexander Parkes found that cellulose in the form of woodflour or cotton fibre, when dissolved in nitric and sulphuric acids and mixed with oils, formed a dough which dried to look like ivory or horn. The trouble was that cellulose nitrate (pyroxylin) is extremely explosive; guncotton is the same chemical substance. The prolific Parkes registered twenty patents for mouldable pyroxylin materials alone. It is worth reading his often-quoted, highly prophetic paper given to the Royal Society of Arts in December 1865, in which he describes his search for a new material capable of replacing all previously known natural plastic substances such as ivory, tortoiseshell, horn, rubber and gutta percha.

Parkesine mouldings, made from 1855 to 1868, are brittle and very rare; they are considered to be the oldest objects in the world made from a man-made material. The best collection is displayed in the Plastics and Rubber Institute in London. Book covers, bracelets, handles and other mouldings are delicately inlaid with precious metals or mother-of-pearl, or painted. Parkesine was also moulded into more mundane objects such as shoehorns, door handles and fishing-reels.

Parkes lost interest and the Parkesine Company went into liquidation in 1868; but the search for a less brittle, mouldable substitute for ivory went on. In 1869 a competition was held in America to find such a material, and this spurred on the Hyatt brothers, John Wesley and Isaiah, to find the solution to the problem of brittleness which had defeated Parkes. An admixture of camphor proved to be the answer, and the Hyatts patented

Celluloid that same year. Celluloid, like Polythene, is a trade name which has passed into common usage; through the association with the cinema it has almost become synonymous with 'unreal'.

Celluloid is the classic semi-synthetic plastic for collectors. Secret manufacturing formulas passed down from worker to worker – as at the Xylonite factory (the main British trade name for celluloid) at Hale End, North London – created patterns and colours never bettered in any plastics material. Delicate tortoiseshell effects could be achieved in any colour; pink coral looks just like the real thing; and lustrous 'pearls' were made by the addition of particles of lead phosphate. Boxes made by hand and from cut and glued sheet sometimes lack the finesse of a finely machined product, but every piece reveals differences almost equalling the irregularities found in natural materials such as wood. In fact, many of the imitation tortoiseshell markings were painted by hand with aniline brown solution mixed with a little fuchsin.

Technically, it is easy to inlay celluloid with precious metals, and monogrammed cigarette cases and book covers can often be found. As with other thermoplastic materials, all waste celluloid can be softened and re-used, and if pressed together again can create even more dazzling mixtures of colours and patterns.

Around 1920, long hair was still the fashion for women; by this time celluloid was taking over from horn and tortoiseshell. A hand-cut celluloid comb came to be more valued than a comb made from any other material. It was said that each woman wore between 100 and 200 grams (about a quarter of a pound) of celluloid in her hair: a grim thought in the light of the flammability of the material. Like the hemline, hairstyles shortened in the 1920s, bringing to an end the era of the giant comb. Even so, in 1944 nearly 90 per cent of toilet goods were still made from celluloid. Many of the larger pieces have a moulded shape cemented round a wooden core; hand mirrors use a hollow frame, blow moulded from two sheets of celluloid.

Other plastics based on cellulose include Cellophane, cellulose acetate and viscose rayon; but the only semi-synthetic rival to celluloid, from the collector's point of view, is casein. Casein is made from skimmed milk curdled with lactic acid; it was patented as a moulding compound in the 1890s by Spitteler in Bavaria. Cured by long immersion in formaldehyde, it became a useful thermosetting material which he first described as 'artificial horn' and registered under the trade name Galalith (milk stone). Casein can look very like celluloid and can also be moulded in extraordinarily beautiful patterns to which there seems to be no limit. It is therefore important for collectors to be able to identify it, and a simple method is described in the final section of this book.

Casein absorbs water easily and starts cracking, but even today it still has a place as a strong competitor to polyester and nylon in the manufacture of buttons.

First synthetics

In the 1900s a Belgian-born chemist, Leo Baekeland, had been searching for a clear, glass-like plastic, but instead had become distracted by the chemistry of phenol (carbolic acid) and formaldehyde, which had been baffling chemists for some time. In 1907, through a series of complicated experiments, he succeeded in converting phenol formaldehyde into the first truly synthetic plastic, the phenolic resin universally known as Bakelite.

Bakelite is a trade name which now covers a variety of plastics, but the word is inseparable in collectors' minds from Baekeland's famous phenolic resin. As this is hard and brittle, reinforcement of fibres or woodflour had to be added, and the mottled effect characteristic of phenolic mouldings is caused by this filler, which also restricts the colours to black and dark shades of brown, green, blue and red.

This original Bakelite is found particularly in connection with 1920s and 1930s radios, electrical goods and cars, where its talent for imitating timber, and its excellent insulating properties, were exploited for moulding radio cabinets, hairdryers, dashboards, knobs and ashtrays. Of all plastics, this is the favourite among collectors, perhaps because it is easy to find, survives the rough conditions of street markets, and was the first plastic to inspire a new and distinctive range of forms. A conscious effort was being made in the 1920s, as Jeffrey Meikle points out in his *Twentieth Century Limited* (1979), to move plastic 'beyond a limited novelty market' by publicizing 'the material's inherent beauty': 'Plastic and industrial design enjoyed a symbiotic relationship.'

If phenolic resin is made without any filler, a very attractive, light, transparent, though brittle plastic is produced which can be used for purely decorative designs. This cast phenolic, in onyx, jade, marble and amber, is easy to machine and was very popular for Art Deco light fittings, candlesticks, fluted boxes and co-ordinated ranges of jewellery, and later for small American radio cabinets. It is very light-sensitive and tends to darken with exposure, so that it is no use for pale-coloured mouldings.

Until the 1920s the discovery of each new plastic was very much an empirical affair, and often an important substance was stumbled upon by accident as the chemist searched for something else. But in 1922 a German chemist, Hermann Staudinger, synthesized rubber, which not only gave the German rubber industry a much needed boost but, more importantly, completely changed the current theories about organic chemistry.

It was believed that plastics were composed of rings of linked molecules, but Staudinger demonstrated, through natural materials which have the ability to be moulded, that they are in fact constructed from huge chains of thousands of molecules linked together. Man-made plastics are made with chains which are as much as 10,000 times longer than natural molecules, and are known as 'superpolymers'. It took Staudinger many years of campaigning to convince his stodgy fellow scientists of this fact, and it was not until the 1930s that this was reflected in a flurry of activity and the appearance of many new plastics.

The main light-coloured plastic previously available, celluloid, was basically a sheet material. However, in 1924, a British chemist, Edmund Rossiter, combined two common gases, carbon dioxide and ammonia, with formaldehyde and produced the first water-white, transparent moulding powder. Urea thiourea formaldehyde sounds unpleasant, but was a great success as tableware under trade names such as Bandalasta and Linga-Longa.

The most beautiful effects of thiourea are undoubtedly the patterns of marble, alabaster and stone which were achieved by either sprinkling coloured powder in a particular pattern round the base of the mould or adding it to the mix beforehand. British Cyanides unveiled thiourea at the 1925 Empire Exhibition at Wembley, and by 1926 a display in Harrods attracted such a jam of curious customers that other department stores followed suit. The event at Harrods saved British Cyanides from bankruptcy, and by 1929 their moulders, Streetly Manufacturing, was the largest plastic moulding company in Britain.

After 1932 an improved material, urea formaldehyde, supplanted thiourea, and the mottled colours of Bandalasta went out of favour. Homes in the 1930s were filled with brightly coloured, compression moulded eggcups and cruet sets, light fittings, cream makers and, of course, picnic sets. An attractive light-coloured domestic plastic had at last arrived, and it was capable of withstanding most kitchen use.

A major reason for the increase in plastics in the 1930s was that oil began to usurp coal's position as the source of the chemicals. This not only brought down the price of moulded end-products, but led directly to the development of new polymers. In addition, problems with the injection moulding machine were resolved; this was to have an important influence on design and ultimately on the quality and shape of our lives. The first commercial injection moulding machine was patented in Germany in 1926, but it was not until 1937 that it became capable of full automatic production, sweeping us into a new era of mass-production and cheaper plastics for all.

The shift away from traditional craft values towards a new machine-based aesthetic encouraged a smart, functional, 'modern' lifestyle. Colourful, hygienic surfaces kept dirt at bay; middle-class housewives were losing domestic staff and wanted to rush off to cocktail parties. The dressing table became a plastics repository, both in toilet articles and the packaging of cosmetics. Smoking became more glamorous, which meant more cigarette boxes and ashtrays; plastics were used increasingly in sales promotion, and here the cigarette companies led the field with some elaborate designs.

Plastics did not manage to hold on to their luxury image, however: in 1939 *Plastics* magazine began to complain of 'cheap, low-grade moulding powders, skimped designs and faulty moulding techniques', and of the 'wild scramble to exploit every available, and sometimes unsuitable, market'. Through no fault of their own, plastics began to be associated with a cheap and nasty image. Plastics manufacturers fought back, and the British Plastics Federation even tried to set up a hallmarking scheme. But in that year war was declared in Europe, and the whole picture changed. Plastics became a vital resource, and revolutionary new materials came into quantity production.

Plastics everywhere

During World War II, in July 1943, John Gloag read a paper at the Royal Society of Arts in London entitled 'The Influence of Plastics on Design', in which he described the new consumer society: 'For the first time in history thousands of women, uprooted from their homes, served in auxiliary forces on a national scale, and came into contact with modern machinery used in the war effort, machinery designed to perform with the utmost economy of effort and material.'

He went on to suggest that the outcome of this experience after the war would be women who were design-aware, critical, and who would not be prepared to put up with badly designed, cheaply made products. Design awareness had already placed mundane plastic kitchen tools in the Museum of Modern Art (New York 1940), when the 'Useful Objects Under $10.00' exhibition presented utensils in acrylic, a wartime plastic. An important aim of the 'Britain Can Make It' exhibition held in London in 1946 was to stimulate industry by showing the direct application of war technology to civilian products. Plastics and rubber featured in a very dramatic display. The plastics industry was to expand more than any other industry over the next four decades, stimulated by the depletion of natural materials and undeterred by post-war austerity.

The most important plastics to emerge after the war were PVC (vinyl), melamine, polyethylene, polystyrene and nylon. Styrene in particular had been overproduced. Nylon was the first totally man-made fibre, discovered in 1938 in the Du Pont laboratories in America by Wallace Carothers. It was swiftly made into bristles and brushes under the trade name 'Exton', while nylons became a valuable fashion accessory. Nylon was even used as a hair lacquer! War-surplus nylon inspired all manner of uses. 'The latest idea for coupon-free curtains and cushions is printed nylon camouflage parachutes. The parachutes come in a pattern of pale and dark green, some in a large design, some in a small. They drape beautifully and are more interesting than in the plain colours.' (London *Daily Graphic,* Monday, 7 July 1947.)

Throughout the 1940s decorative laminates made great progress in America, and Formica, Roanoid and Micarta surfaces beautified kitchens, trains, airliners, hotel bars and cinemas. Fire-resistant and very tough, melamine-faced laminates went on to characterize the 1950s era of dinettes and espresso bars.

Formica was such a practical material that it soon became the standard surface in cafés and hotels. Maybe the inhabitants of the espresso bars of the late 1950s and early 1960s were really the first true plastics generation, with their Terylene shirts, moulded (pleated) polyester skirts, beehive hairdos sealed in position with vinyl acetate lacquer, and legs encased in 'sheerer era' nylon stockings suspended from synthetic rubber Lycra roll-ons. All these materials appeared in the 1950s. And would the Hula-Hoop madness which created a market for 60 million plastic rings by the end of the decade have occurred without war-surplus Polythene?

It is virtually impossible to salvage historical examples of decorative laminates, but moulded melamine tableware produced some valuable designs, many of which are still in everyday use. As a competitor to china, it offered much richer colours as well as being unbreakable, but it was more expensive. America led Europe, and by the late 1950s as much as 50 per cent of all dinnerware sold was moulded in melamine. Russel Wright's designs were particularly stylish, while in England there was Ron Brookes working for Brookes and Adams.

Polyethylene is a polymer of ethylene, derived from oil, and had been discovered by accident in 1933. During the war it was polymerized at high pressure into low density polyethylene, and its widespread appearance in the 1950s was the result of a method that was developed in Germany in 1953 for polymerizing it at much lower and safer pressures. The new high-density polyethylene – its Du Pont trade name, Polythene, is the name that has caught on, and is used as a generic name in Britain – was stronger and harder, with a higher melting-point, and began to invade areas where previous plastics had failed: buckets for hot water, sturdy dustbins, baby baths and containers for all kinds of chemicals.

Until polyethylene celebrated its fiftieth birthday in 1983, it had been sadly neglected by collectors, usually ending up in a Polythene trash can. Now its most famous product, Tupperware, is coming into its own. The Tupperware Corporation was the first firm to make a positive effort to move this particular plastic up-market, in keeping with the mood of prosperous 1950s consumerism, and Tupperware food containers are prized for their design as well as for the fascination of the photographs of early Tupperware Parties.

Polythene is the most ubiquitous plastics material manufactured, one of the most 'plasticky' plastics, and probably, because of its universality, the most despised. Yet its reliability and versatility were important factors in establishing public acceptance of plastics in general. In the 1980s the appearance of a new version, linear low-density polyethylene (LLDPE), cheaper to produce than the old low-density material, perhaps heralds a new crop of innovative designs.

The 1950s also brought us PVC, a plastic that suffered in the beginning from instability. It was stronger than Polythene, and among its countless applications is the making of records, for which it ousted shellac and phenolic. Long-playing records and 45 rpm singles first appeared in 1952.

High technology

The 1960s were good times for plastics, and industry was generous in investing in the development of patentable polymers. Business was so successful, in fact, that the problems of the recession-hit 1980s, when factories have an over-capacity for producing synthetics, stem from these boom years.

Technical progress released a wide range of plastics which were eagerly grabbed by designers, impatient for the media with which to express the new, informal, iconoclastic lifestyle. Expanded polyurethane foam, both flexible and rigid, freed furniture from the restriction of traditional construction, and furniture students had to study the structure of laminated foams of different densities. They could push bodies around at will, on and into what was optimistically described as 'fun foam' furniture. Abstract forms were linked together in endless variations, such as Jørn Utzon's puzzle-like seating, or the hinged geometric system designed by Johannes Larsen for France & Son, Denmark. Foam chairs sprang up from prisons of flat, compressed envelopes; or the foam was simply wrapped around a sculptural metal frame. The development of soft and hard foams, with a simultaneously moulded protective skin, resulted in a rash of unusual, eccentric forms, such as the shapes produced by Gufram in 1969-70. Self-skinned foams masqueraded as grossly oversized blades of grass, prickly foam cactus hat stands, and boulders to be arranged casually in the home or garden.

The sharp, shiny image of the 1960s was created entirely by synthetic plastics: wet-look polyurethane, glossy ABS, shiny, transparent acrylic and PVC. The aesthetic invisibility of acrylic actually obscures the fact that it is quite impractical: it attracts dust beautifully, it scratches, and it chips when dropped. Inflatable PVC furniture too was liable to suffer punctures from the odd cigarette or sharp object, though a puncture patch was never to be seen spoiling the purity of the designs.

Layers of coloured acrylic can easily be laminated to create refracted patterns, and this was used in bright, psychedelic-style clocks and jewellery. Large objects could now be blow-moulded, and the Italians shaped hundreds of light fittings, exploiting the translucency of acrylic. The 'Prospex 67' exhibition sponsored by ICI at the Royal College of Art, London, in 1967, was a significant promotional event, as it gave students the chance of using ICI's Perspex acrylic in any way they fancied. Dunlop also promoted their soft foam rubber by instigating a biennial, and later annual, competition.

The ease with which plastics insinuated themselves into everyday life misled the consumer into believing himself to be at the centre of a disposable world. In reality, these materials were very expensive products of high technology. Inflatable furniture eventually deflated, and foam, even sprayed with a PVC skin for a longer life, did not survive.

By the late 1960s plastics had made technically possible the destruction of our familiar domestic world. The important and aptly titled exhibition, 'Italy: The New Domestic Landscape', staged in New York in 1972, proved to be the visual and polemical summary of the crisis in design. The catalogue encapsulated in a tracing-paper cover cut-outs of plastics products, which shifted loosely around inside.

On the subject of plastics, Giulio Carlo Argan, professor of history of art, describes the products of the Italian firm Kartell as 'exemplary'. Their pieces 'are constantly changing . . . The shapes are not governed by any fixed or precise aesthetic notions but serve as typical signs. The "thing", besides not being of any intrinsic worth, and therefore not requiring preservation indefinitely [except now by collectors and museums (author)], is also resilient and almost transparent. It has no existence of its own but is merely a transitory presence in the course of one's life.'

Plastics had been set free and were poised for flight into the future. Unfortunately the 1970s turned out to be a disastrous decade for the plastics industry. 'Cynical sophistication' was Philippe Garner's phrase for the 1970s, and the plastics industry certainly had something to be cynical about. Firstly there was the energy crisis of 1973, which resulted in shortages of materials, and competition over petrochemical feedstocks. This then led to inflation and recession, and by 1979 oil prices had actually quadrupled, which meant belt-tightening, cutbacks, and a halt to hopes of expansion.

In spite, or maybe because, of this the accent fell on exploiting the plastics that were there. Almost without the public realizing it, the newer superpolymers, plastics with engineering properties, began slowly to replace metals. Miniaturization, too, brought about by microchip technology, created outlets for plastics in hundreds of housings and cases for all kinds of equipment. The domestic landscape started to reflect the industrial aesthetic of 'High-Tech', the factory was brought into the home, and the barriers blurred.

No one was in the mood for much frivolity, and in reaction to the synthetic 1960s, there occurred a revival of the traditional values of craft, of ethnic designs, of wood, leather, metal and feather. Carved foam blocks and air-filled cushions had not always been ergonomically suited to the human frame, and the hippies were growing older. The nostalgia craze of the late 1960s had not left much space for plastics. And everything had to be vandal-proof.

Finally, towards the end of the 1970s, a spark of hope illuminated the streets, the eruption of British Punk. Tacky, despised, colourful plastics – vinyl in particular – were ideal for expressing the aggressive reaction towards a society that appeared to be filled more and more with plastic people. Singers like Poly Styrene and record albums such as Plastic Letters exaggerated the 'plasticity' and emptiness of life, and even beyond the Iron Curtain a group with the name of Plastic People of the Universe took up the theme. Plastics were useful, not just as body ornament, abstract and inventive, but in maintaining the Mohican and other startling new Punk hairstyles. Hair gel and setting lotion are made of a type of PVC, polyvinyl pyrolidone.

The functionalist reaction of the early 1970s had all but annihilated decoration and humour from the grammar of design. Now again in the 1980s a new ornamentalism has surfaced, notably in architecture, furniture and domestic fittings, but also reflected in fashion, beginning with the decoration of the body, sprayed with colours, with the carefully ripped and pinned-together clothes and jewellery using plastics in new, inventive and skilful ways. Designs are becoming more abstract, but in essence decoration has become totally separated from function, and is a priority set above function.

It is an ideal situation for plastics, being materials of alchemy, adaption and camouflage. Old-fashioned decorative laminates have not exactly inspired designers during the past fifteen years or so, but new types of laminate from America and Italy have inspired a revival of products surfaced with plastics.

In the early 1980s the work of Studio Alchymia in Italy pioneered the resurgence of laminates. Ettore Sottsass, a member of the group, went on to form his own offshoot company in 1981, Memphis, which produced at record speed something very different and definitely plastic for the Milan Furniture Show that year. Coffee tables and sideboards, cabinet-bars and seating in quasi-geometrical forms, borrowed motifs and materials from past eras but assembled them in totally new forms. Described as 'The New International Design', the work is characterized by a riotous use of multicoloured plastics laminates. Primarily selected for surface colour and pattern, these laminates appear to lend the pieces an air of common practicality, even though the bookcases are sloping and the seats are horribly flat. Memphis designs are presented each year like a fashion collection, and the prices are similarly outrageous. £2,000 or $3,000 for the Carlton sideboard by Sottsass places the furniture squarely in the collectors' market for limited editions.

The confused customer, suspicious of being taken advantage of, finds himself in a cultural dilemma, the Emperor's New Clothes syndrome, and journalists have agonized over it. Is this all tasteless kitsch, or is it OK because some people take it seriously? 'This is a period of modifications,' announced Alessandro Mendini. 'We are creating new metaphors and utopias.' The Alchymia pieces use a new high pressure plastics sheet material produced by Abet Laminati, which commissioned several of the Memphis designers to produce silkscreened patterns for the range, and which in 1983 Abet was bold enough to advertise as 'Gli oggetti del paesaggio artificiale' ('The objects of the artificial landscape').

The Formica Corporation in America has also entered a new era with another material, Colorcore, which is not a laminate in the sense of a surface material to be veneered over a basic structure, but a solid plastics medium in its own right. True, it is made from layers of kraft paper impregnated with resin as in standard laminates, but the expensive melamine continues all the way through, and is not just limited to the top surface. This produces a material which is harder than some hardwoods and one which can be machined on standard workshop machinery; because the colour goes all the way through, the material can be routed, sandblasted and bevelled in a way which only used to be possible with cast plastics such as acrylic. Several colours laminated together produce a vibrant edge decoration. The 'Surface and Ornament' exhibition held in Chicago in June 1983 was the first public viewing of Colorcore, and there is no doubt that this exciting, innovative material will produce entirely new concepts in design.

Other designers are using plastics to free products from traditional limitations, such as Daniel Weil whose quirky radios hang about in transparent PVC envelopes, like large, decorative fly traps. The 'Objects for the Electronic Age', designed by George G. Sowden and Nathalie du Pasquier, appear to span the divide between the world of frightening new images and the restrained, familiar forms of industrial products.

In Italy and West Germany acceptance of plastics has always been advanced, and industrial designers are known for their cool, elegant designs. Castelli's furniture for Kartell is never uninteresting. Super-minimalist Dieter Rams, head of Braun product design, represents the extreme state of the form-follows-function credo in his search for perfect performance, pruning any ornamentation that dares to stick its nose out at him. His designs have been collected by classic-hunters and have been exhibited for over twenty years in museums.

As plastics stride ahead continually improving their properties, a spin-off has been a positive move towards sophistication in the alternative world of fakes and substitutes. The 'Ramboo' rattan, bamboo and canework moulded in Mexico by Plastirama astounds us with the quality of its reproduction, and one is forced to concede a certain beauty. To add to the confusion, the manufacturer states that these PVC mouldings have a 'better natural appearance' and are 'better finished' than the real thing! Ironically, although the PVC is tough and weatherproof, a 'wood wax' has to be applied. Companies such as Fashion Foliage have raised the imitative to a growing, popular art. The flame-resistant, easy-care polyester silk palms, ficus, cyclamen and dieffenbachia are sometimes too real for comfort. In 1981 a thriving business in imported fake suede and leather coats was exposed in London when they began to flake after dry cleaning. The PVC was utterly convincing and the coats had been manufactured to the highest standards.

Our attitudes have been forced to change. From being accustomed to plastics bowls or clothes that melt when they come anywhere near heat, we must now adapt to plastics which completely replace metal: kettles, curling tongs, boil-in-the-bag dinners, even a plastics 'tin can'. A plastics automobile engine is being tested in America. Artificial hearts make the biggest news stories, but other medical implants and instruments are quietly saving countless lives.

What is certain in the 1980s is that there will be more and more emphasis on improving the properties of existing plastics so that they can replace other materials to greater advantage. Collectors should take note of these take-overs as they happen. Where previously expensive research projects would have been funded, now step-by-step developments such as blending of existing plastics, or progress in coextrusion technology, will produce new forms and new functions.

Snuff box
Pressed horn
18th century

Horn is a natural thermoplastic material, shaped like most later plastics by heat and pressure. The motto is *In nocte laetamur*: 'We revel by night.'

1 The birth of plastics

Inventors and craftsmen: natural and semi-synthetic plastics from horn to celluloid

The box, above left, is made of a natural plastic material. Lacquer, a resin tapped from trees, is built up in thin layers, each of which hardens (polymerizes) on exposure to the air. The camellia design is carved into the top. The simpler and quicker press moulding technique made the brooches (above right) from powdered horn and from the rare bois durci, woodflour bound with animal albumen.

The most valuable of the early semi-synthetic plastics is Parkesine, from which the objects (right) were moulded. They illustrate the colour effects made possible by the incomplete mixing of pigments, and the use of an intricate piqué inlay of gold, silver and mother-of-pearl. It is only recently that makers have begun to recapture some of the precious, decorative quality of early plastics such as these.

Box with camellia design on the lid
Carved Chinese lacquer
China, early 15th century

Brooches
Horn, with bois durci brooch in the centre
England, 1880s

Mosaic samples, knife handles and brooch
Parkesine cellulose nitrate
Made by Alexander Parkes
England, 1860s

Alexander Parkes, founder of the modern plastics industry
Painting by Abraham Wivell, Jr.

Plaque of cupids with ram
Parkesine cellulose nitrate
Made by Alexander Parkes
England, 1860s

Letterpress
Moulded Parkesine cellulose nitrate
Made by Alexander Parkes
England, 1860s

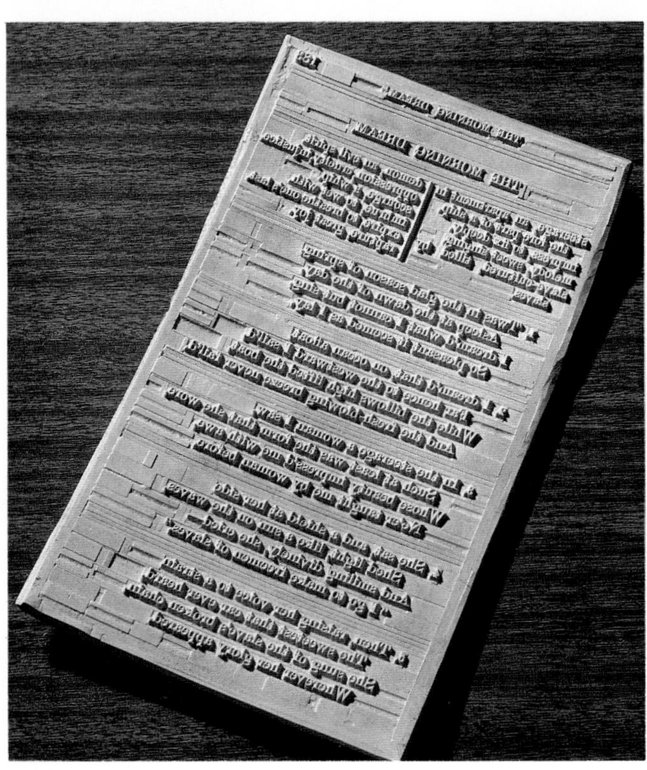

Display of indiarubber boat, pontoons and other rubber articles at the Great Exhibition, London, 1851
Made by Charles Goodyear
USA

Native rubber moulding of head
Unvulcanized natural rubber
South America, exhibited London, 1851

To find and identify a Parkesine moulding is every collector's dream. Alexander Parkes never produced on a large scale, and went out of business before the similar but less brittle celluloid was developed. The plaque (opposite page) indicates the way in which, in plastics, a basic pressing can be refined by hand-carving to create undercuts in high relief.

At the Great Exhibition of 1851 in London, Charles Goodyear showed a number of inflatable pontoons, boats and buoys in fabric coated with unvulcanized natural rubber (latex or caoutchouc). 'The material of which they are formed', he claimed in the catalogue of the exhibition, 'precludes the probability of injury from concussion.'

Union case
Shellac
Moulded by Holmes, Booth & Haydens
USA, 1860s

The earliest plastics mouldings found by the antique hunter are likely to be ornate, black union cases (carrying cases for early photographs) or dressing-table articles in shellac. Like lacquer, this is a natural thermoplastic resin, but of animal origin. From the 1850s onwards it was mixed with filler and pressed under heat to make intricate mouldings; similar pieces were later made from vulcanite and even black celluloid.

Hairbrush and hand mirrors
Shellac
England, 1860s

Chair
Papier mâché inlaid with mother-of-pearl
England, 1840

One of a pair of desk tidies or letter pockets
Lacquered papier mâché
Probably England, 19th century

Page from the Gutta Percha Co. catalogue
England, 1850s

Papier mâché was used by the Victorians as a decorative substitute for wood.

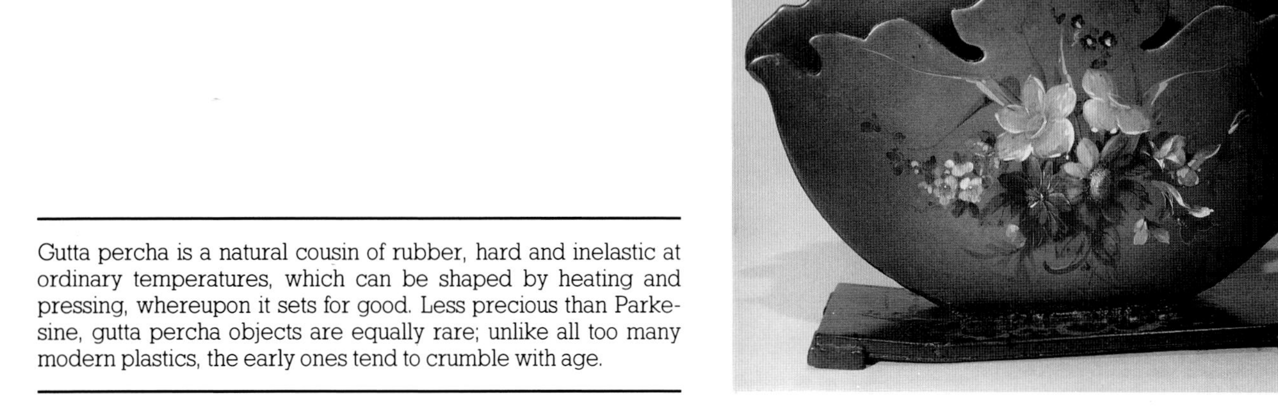

Gutta percha is a natural cousin of rubber, hard and inelastic at ordinary temperatures, which can be shaped by heating and pressing, whereupon it sets for good. Less precious than Parkesine, gutta percha objects are equally rare; unlike all too many modern plastics, the early ones tend to crumble with age.

GUTTA PERCHA BASKETS AND VASES, &c.

Work Basket.

Fancy Basket.

No. 8, Vase.

No. 2, Vase, with Cushion.

No. 7, Vase.

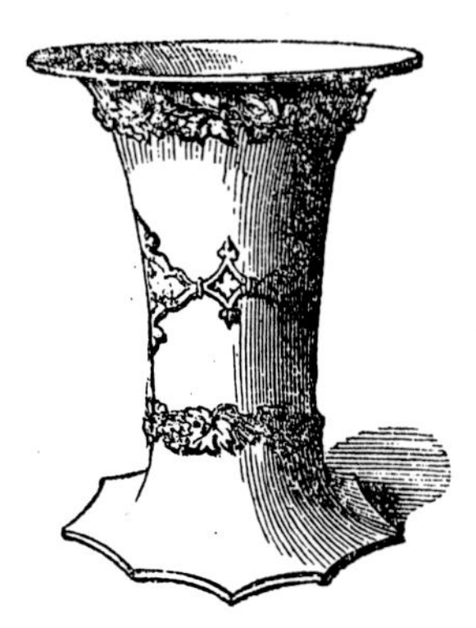
Lighter Stand, No. 5.

Fox's Head Drinking Cup.

Vase, No. 6, (various Patterns)

Ornate inkstand
Gutta percha
The Gutta Percha Co., now Telcon Plastics
England, *c.* 1851

Vesta matchboxes depicting a Jubilee portrait of Queen Victoria, and a Coronation portrait of King Edward VII
Ebonite
England, *c.* 1901

Medical dropper marked 'Puritas', box with French coin embedded in the lid, and two chatelaines
Ebonite
France and England, late 19th century

The face on the ornate gutta percha inkstand has a tale to tell: it is believed to be that of Neptune, moulded to commemorate the successful use of the material to insulate the world's first submarine telegraph cable, from England to France, in 1850.

All moulded rubber used nowadays is vulcanized: heated with sulphur in a process invented by Charles Goodyear in 1838. Carried to an extreme, this process results in hard rubber, otherwise known as ebonite or vulcanite. The Victorians loved jet, and ebonite was used in large quantities as a substitute (see p. 128 for a modern revival); it looks convincing, but is betrayed by its whiff of sulphur.

Comb
Carved horn sculpted in the form of flowers, with miniature pearls
Attributed to Lucien Gaillard
France, *c.* 1900

Hair ornament
Horn carved in the form of two sprays of cow parsley, studded with seed pearls
Designed by Lucien Gaillard
France, *c.* 1900

Comb
Carved horn decorated with stems and leaves of gold and set with small pearls
Designed by Henri Téterger *fils*
France, *c.* 1900-05

Doll
Celluloid and kid
Manufactured by Rheinische Gummi- und Celluloid-Fabrik
Germany, *c.* 1910

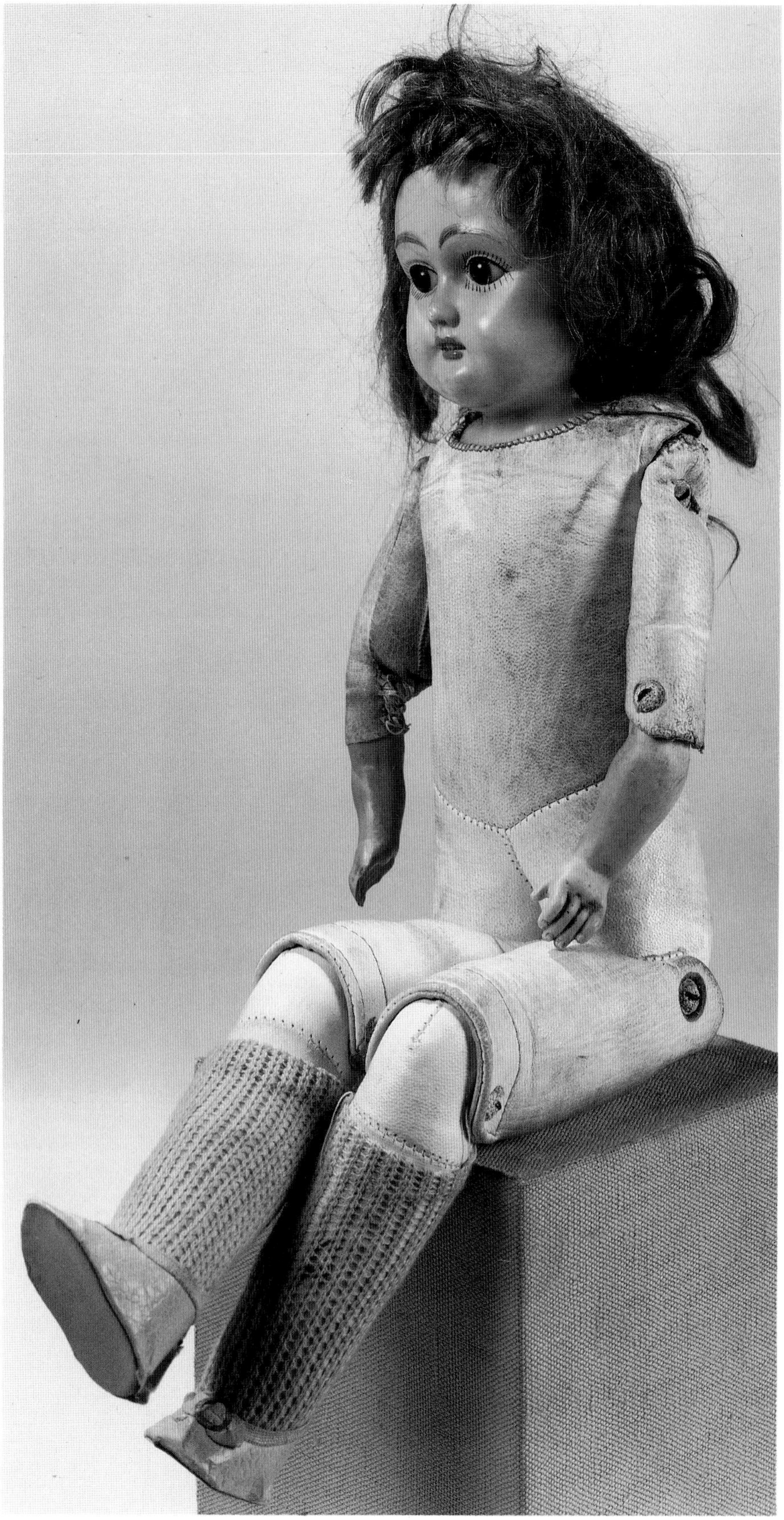

Although a natural polymer, and a thermoplastic material which can be moulded by heat and pressure, horn is the traditional material for combs and hair ornaments cut by hand, as in the delicate Art Nouveau examples. The curved comb has been heat-shaped before carving.

The doll (right) has her exposed areas moulded in tinted celluloid, the commercial successor to Parkesine.

Dressing table tray with nail polish boxes, cut-throat razor, glove stretcher, nailcare tools and photographic tray
Celluloid
England, late 19th century

Onoto and Swan fountain pens with their boxes
Ebonite
Manufactured by Thomas De La Rue, London, and Mabie Todd, Toronto
England and Canada, 1880s-90s

Celluloid, often in simulation of bone, ivory or tortoiseshell, soon found its role in the manufacture of light household objects, alongside ebonite. Slender ebonite fountain pens, often somewhat faded now, are among the most popular of plastics antiques.

THE "ONOTO" PEN
SELF-FILLING
SAFETY FOUNTAIN
PATENT

The SWAN FOUNTAIN PEN
MABIE TODD & Co.
Size 3
TRADE MARK

SAFETY SCREW CAP

Plastics from milk: casein is a material in which almost any colour or pattern is possible. The great Art Nouveau designers, with their love of the precious and their disdain for the merely valuable, took delight in using it alongside ivory and other inlays.

Mantel clock
Stained sycamore inlaid with metal, abalone and mottled, iridescent casein and with clock hands of copper
Designed by Josef Olbrich
Austria, 1902

Mantel clock
Ebonized wood inlaid with casein, ivory face
Designed by C.R. Mackintosh
Scotland, *c.* 1919

Ornate comb, hairpin box and wrist bag
Celluloid
Late 19th century

Fan
'Tortoiseshell' celluloid sheet with silk ribbon
England, late 19th century

Celluloid used for its beauty: the comb is hand-cut and heat-shaped, and the moulded bag-clasp and chain distantly recall ivory. The translucent box, with clear patches probably imitating horn, is moulded round a reinforcing wire rim. The fan segments are cut and trimmed exactly like a horn or tortoiseshell fan. The material may be cheaper, but the craftsmanship is essentially the same.

Salad servers, powder box, cylindrical hairpin box, shoe horn and pill boxes
'Ivory' celluloid, heat formed or blow moulded
England, 1890s

Babies' rattles
Blow moulded celluloid sheet glued to heat-bent celluloid tube
Probably France, late 19th century

The moulded salad servers (both from the same mould) and the shoehorn have much of the weight and feel of ivory. They have been moulded from 'ivory' Xylonite celluloid sheet without the tell-tale parallel lines left by the planing machine on nearly all sheet celluloid. The lids of the feather-light boxes are blow-moulded. Table-tennis balls, made in the same way as the balls in the rattles shown below, are among the few articles still made in celluloid. Like all celluloid (and Parkesine), the rattles are chemically close to guncotton; such explosive toys were still common well into the twentieth century. Their colouring shows the beginning of celluloid's emancipation from the imitation of other materials.

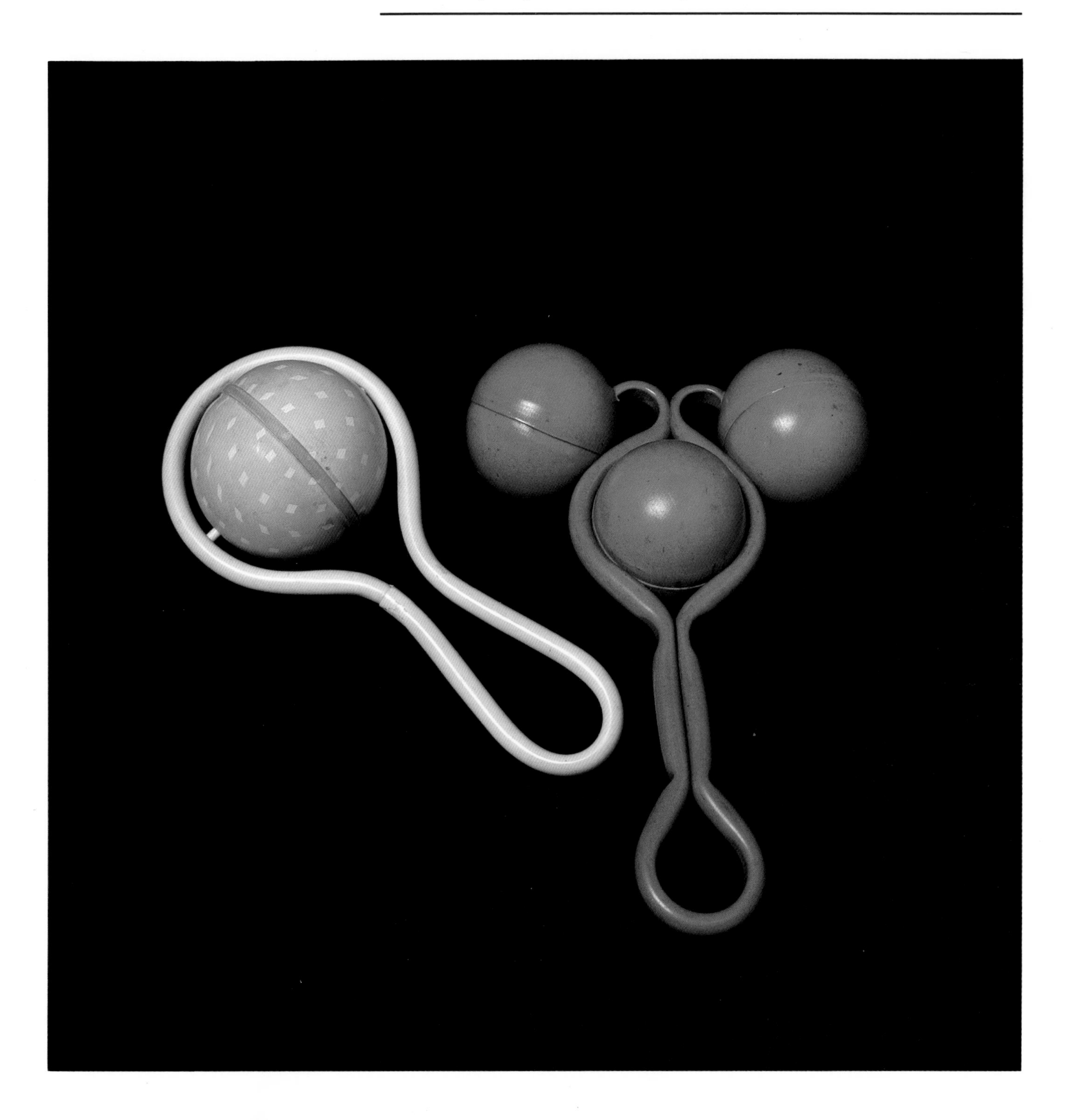

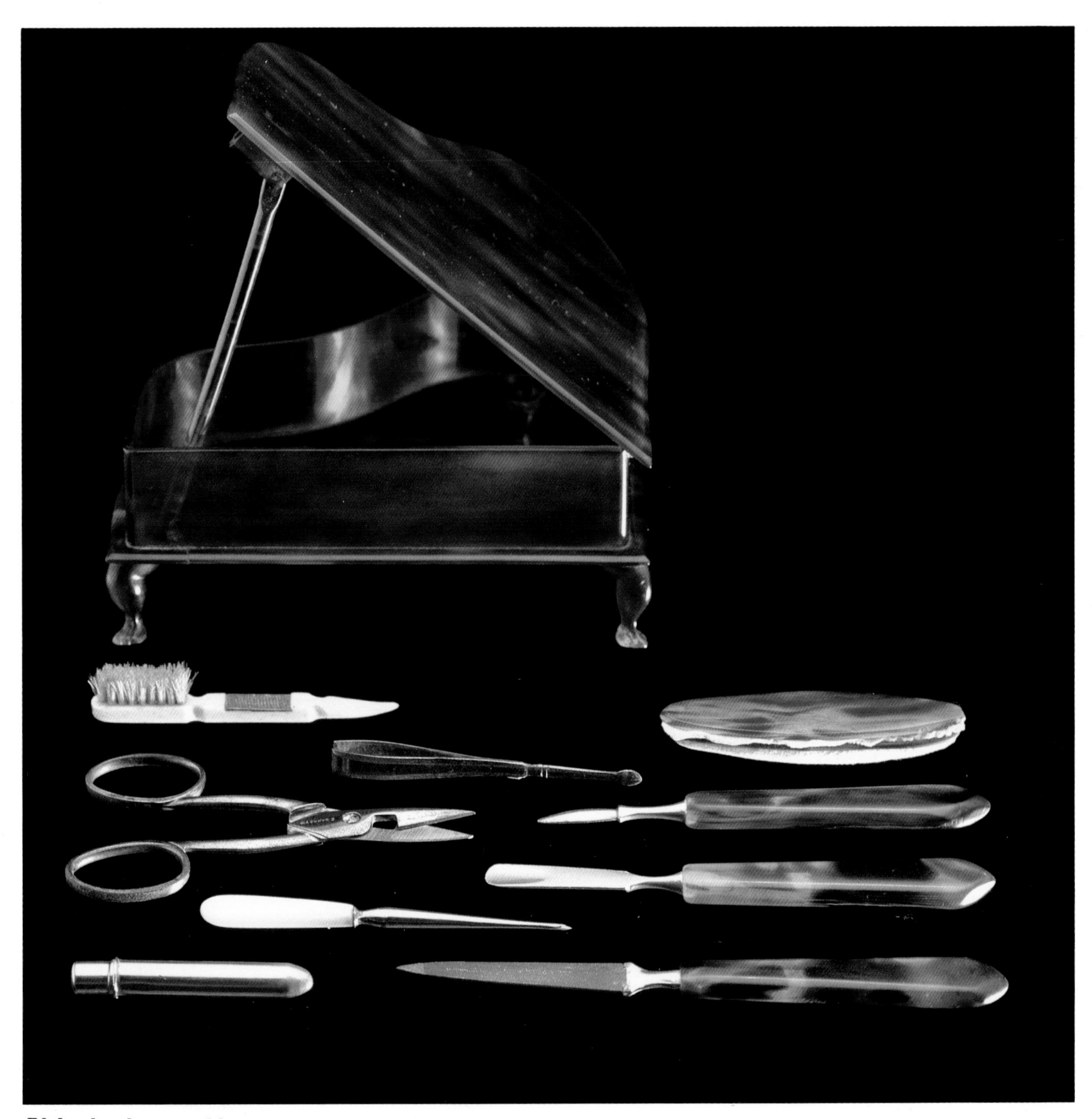

Zéphyr hand-operated fan
Celluloid inlaid with silver and mother-of-pearl piqué
Manufactured by the Zéphyr Company, Paris
France, 1901

Piano-shaped nailcare set
'Tortoiseshell' celluloid; handle of file, in real tortoiseshell, handle of nail parer in casein
England, before 1912

The set of articles in the piano-shaped case appear all to be made of the same material; but burn tests reveal that one piece (bottom right) is tortoiseshell and another (second from top, right) is casein; the rest are celluloid. The combination of plastics with a natural substance is unexpected. The ingenious Zéphyr fan is one of those objects in which celluloid – which has always been a cheap, domestic plastic – is honoured with a precious inlay.

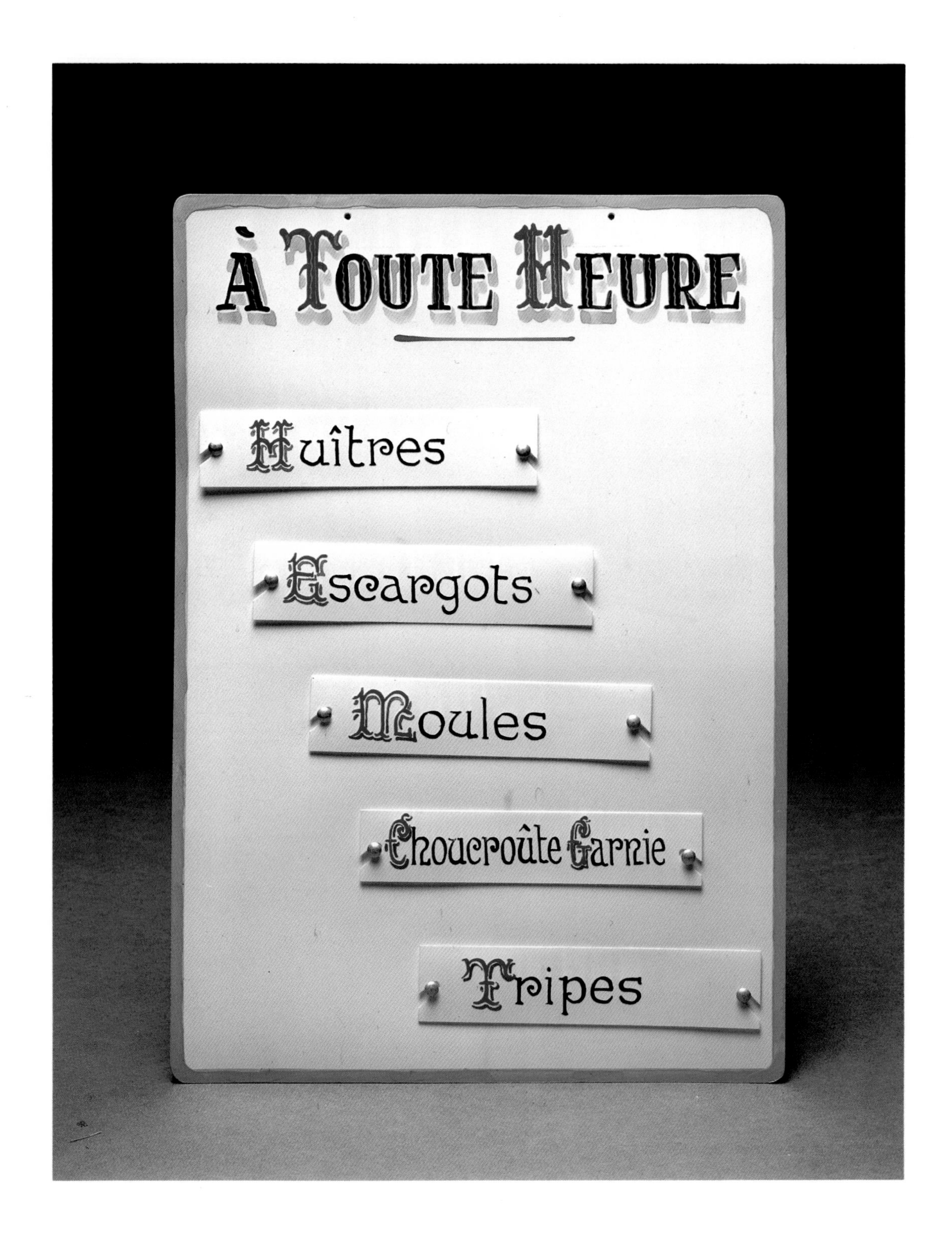

Plastics for convenience: the menu is a traditional French design made for many years, with the dishes of the day slotting over rivets.

Café menu sign
Hand-painted celluloid sheet
France

Bandalasta Ware Tennis Set
Green urea thiourea formaldehyde
Moulded for Brookes & Adams by Streetly Manufacturing
England, 1927-32

2 Artistic and useful
The 1920s

The first synthetics: entirely new colours and shapes appear in the home

The palette-shaped picnic set, a product of 1920s informality in social life, is described as adding 'enjoyment to alfresco meals. For the Car, River, Garden or Ramble'.

Bakelite, unlike celluloid, has never quite shed the capital letter which marks it as a trade name: the word still, in fact, denotes a proprietary range of plastics rather than one material. In users' and collectors' minds it has remained inseparable from the dark, often grainy or mottled, glossy material which has the generic name of filled phenolic resin. Not all phenolic is made by Bakelite, however. The various fillers which give phenolic its strength and insulating properties also often make it difficult to mould in fine detail, and it seems generally to have inspired designers to create emphatic and massive shapes. The Bates index could be a classical architectural bracket. The bomb-like tea-caddy, and the box with its tobacco-leaf edging, both show a typical contrast of textures. The feet of the loudspeaker are graced with the classical patera motif.

Radio speaker
Phenolic
Moulded by Thomas De La Rue for Mullard
England, *c.* 1928

Tobacco box and tea caddy
Phenolic
Probably England, 1920s

Bates Address Index
Phenolic
Designed by Norman Bel Geddes
Manufactured by the Bates Manufacturing Co.
USA, 1924

Two pages from the original Bandalasta catalogue, showing domestic tableware and picnicware
Brookes & Adams
England, 1927

Bandalasta
(M)
(N.G.B.)
(Y.8.B.)
(R9.B.)
(Y.8.)
(M)
(G6)
(R7)
HORN
(H)
(M)
B15
R9
Y8
Marble
G
NY8
N
R7
Y7
(Y8.B.)

Translucent marble and alabaster effects were achieved in a range of designs in urea thiourea formaldehyde, marketed in Britain as Bandalasta Ware. It could also be made in solid colours (see p. 41). Bandalasta was marketed as 'both artistic and useful'; conveniently for plastics historians the trade mark is easy to date, as it disappeared on the arrival of an improved plastic, urea formaldehyde, in 1932, amid signs of a switch of public taste away from marbled patterns.

The characteristic marbled pattern of much plastic ware of the late 1920s is often created by colouring assorted batches of material and grinding them to different degrees of fineness so that they mix imperfectly in the mould.

Bandalasta Ware
Dressing-table powder bowl with lid, sandwich box and dressing-table set
Blue and white mottled urea thiourea formaldehyde
Designed by Arnold Brookes
Moulded for Brookes & Adams by Streetly Manufacturing
England, 1920s-30s

Beakers
Urea formaldehyde
Moulded by Streetly Manufacturing and Stadium
England, 1930s

Even filled phenolic, most familiar in the grainy brown guise popularly associated with the name of Bakelite, can appear in a variety of colours, mottled by its filler and a mixture of pigments.

The beautiful Bandalasta Ware bowl is made in two pieces: the top screws into the base with a threaded brass insert. The piece shown on the right, with its rare combination of solid silver and synthetic resin, was probably made to celebrate the launching of the range. The same design appears in the 1927 catalogue without the silver trim.

Nest of Bandalasta Ware picnic horns
Beatl urea thiourea formaldehyde
Moulded for Brookes & Adams by Streetly Manufacturing
England, 1927-32

Belplastic lidded pots, Shellware beakers and small red games cup
Phenolic
England, 1920s

Bandalasta Ware fruit bowl
Marbled Beatl urea thiourea formaldehyde, with silver rim hallmarked Birmingham 1927
Designed by Arnold Brookes
Moulded for Brookes & Adams by Streetly Manufacturing
England, 1927-32

Bandalasta picnic ware, cruet and powder bowl
Beatl urea thiourea formaldehyde
Moulded for Brookes & Adams by Streetly Manufacturing
England, 1927-32

Master Incolor cocktail shaker
Urea formaldehyde with silver plated cap and recipe rings
Moulded for William & Gill by Thomas De La Rue
England, *c.* 1935

Champagne glasses
Clear polystyrene, injection moulded in two snap-fit parts
Manufactured by Comet
USA, *c.* 1981

3 Mass-produced glamour
The 1930s

Plastics lend themselves to ultra-modern living, and also to flashy design

The upper section of this solid and imposing cocktail shaker shows the familiar stepped 'ziggurat' shape of Art Deco design in a streamlined form. This is directly linked to the nature of the material and to the technique of manufacture: tapering and rounded forms come cleanly out of the mould and resist chipping.

'Evening In Paris' perfume presentation box
Blue mottled phenolic with injection-moulded cellulose acetate shoes
Moulded by Prestware for Bourjois
England, 1939

Fantasy creations of the 1930s: would any of these have been thinkable without plastics? The profile on the right is one of the few entirely non-functional 'art objects' in the book. Its flighty mood contrasts with the sub-Epstein modernism of the bottle opener. As for the little door, which opens to reveal a perfume bottle, *Plastics* magazine claimed in 1939 that 'as long as we can produce ideas such as this in wartime, and so retain our sense of humour, all is not lost with the British people'.

Bottle opener
Injection moulded urea formaldehyde figure coated with metallized cellulose acetate lacquer, metal opener
Manufactured by Alfred Herbert
England, mid 1930s

Lady's head
Cut and engraved clear acrylic sheet with painted lips
Probably USA, 1930s

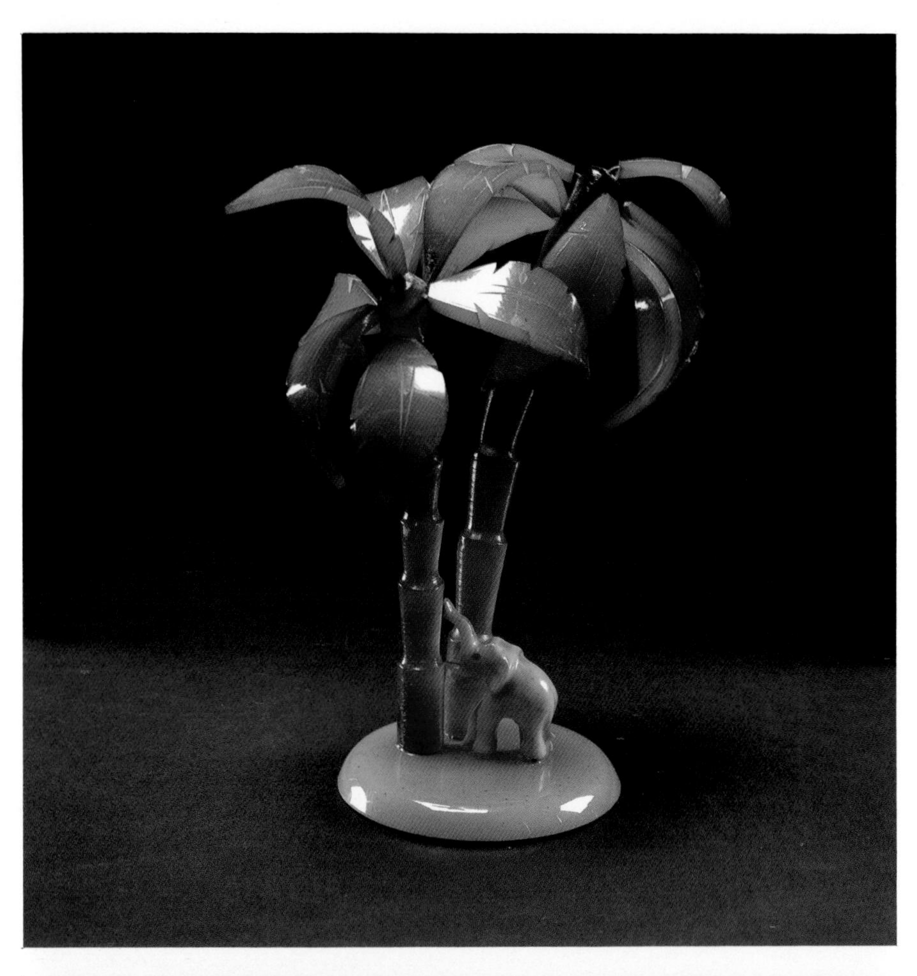

Desert island with two palm trees and elephant
Cast phenolic carved to imitate amber
Carved by Stanislaw Kucharczyk
Poland, 1930s

MUM streamlined sweeper for removing dust, probably from baize gaming tables
Casein sheet glued over wooden blocks
France, 1930s

The little table-sweeper, unlike a lot of streamlined 1930s forms, is entirely functional: as it moves across the surface, static electricity on the grooved base picks up the dust. The tail-fin is a small pull-out bristle brush. The island ornament, on the other hand, is pure fantasy. It is made from unfilled (hence non-structural) phenolic resin, hand carved from cast blanks. The light fitting, moulded in three parts in the utilitarian filled phenolic associated with the name Bakelite, is pure Art Deco. Thousands of similar items were installed in suburban homes; and while many people are collecting them others are probably still throwing them out as junk.

The circular radio is a classic synthesis of function and technology. It could not have been made out of any other material. Practically, the dial is large and legible; symbolically, the whole receiver has become one immense spinning tuning-knob.

Ekco AD 65 radio
Phenolic with celluloid window
Designed by Wells Coates
Manufactured by E. K. Cole
England, 1934

Wall-mounted electric lamp holder
Brown mottled phenolic
Probably England, 1930s

4 COUPS 25c
4 JEUX
TABLE "KIDF"
METTEZ 25c
puis actionnez de droite à gauche
une des manettes qui sont sous la table

The plastics top of the Kipé table contains four games. Drop 25 centimes in the slot, and the games can be activated from below.

Kipé café games table
Phenolic on metal base
France, 1930s

Rolam circular calendar
Phenolic with celluloid window and chromed steel rim
Manufactured by Calendox
England, early 1930s

Superinductance Receiver Type 720A with heptagonal speaker Type 2115
Phenolic
Manufactured by Philips
England, 1931-32

Dienes Reform coffee grinder
Phenolic, designed by Peter Dienes
Manufactured by Pe.De
Germany, *c.* 1924

Filled phenolic became so identified with 'Bakelite' moulded radio cabinets that other objects moulded in the same material, such as the calendar on this page, took on similar forms. Restricted in colour by its filler, and in form by the compression moulding process, it was a material that developed its own vocabulary of strong, simple forms. The housing of the coffee grinder, assembled from several mouldings, is an unusually complicated design.

Beauty became big business in the 1930s, and the industry began to exploit the up-market possibilities of plastics. Reco Capey's designs, for instance, are unmistakably luxury items, as are the cases produced (often as premiums) by cigarette companies. The obvious material was the new urea formaldehyde, made in light colours with fine fillers which allowed delicate detailing.

Powder compact
Urea formaldehyde lid with bee motif in the centre
Designed by Reco Capey
Manufactured by Yardley
England, 1936-38

Powder compact
Urea formaldehyde lid moulded in the form of a flower with a bee in the centre
Designed by Reco Capey
Manufactured by Yardley
England, 1936-38

Two cigarette boxes
Ivory and black urea formaldehyde
Box with Grecian dancers and musicians was designed by A. H. Woodfull for the Ardath Tobacco Co. to hold fifty State Express 555 cigarettes. Moulded by Streetly Manufacturing, England 1935
Larger box 'Elo Ware'. Manufactured by Birkby
England 1936

In the USA, radios became a fashion item. Thousands of small radio cabinets were made from decorative unfilled cast phenolic (Catalin or Marblette). The choice of colours was endless: onyx, marble, jade, coral, rose quartz. The size of these radios reflected the development of smaller components, and the material, unsuitable for larger mouldings, was easy to work on standard equipment into an endless variety of forms. The one above looks like a model of a 1930s cinema.

Radio cabinet
Red and yellow cast phenolic
Manufactured by Addison
USA, *c.* 1934

Baby radio cabinet
Yellow cast phenolic
Manufactured by FADA
USA, *c.* 1934

Radio cabinet
Marbled white and deep blue cast phenolic
Manufactured by Addison
USA, *c.* 1934

Radio cabinet
Yellow and red cast phenolic
Manufactured by Addison
USA, *c.* 1934

70
60
55
Addison

70
60
55
140
160
Addison

Streamlining is an entirely practical shape for an iron; here the handle takes the form of fins integrally moulded with the rest of the housing. The Duplicard, on the other hand, is a collection of 1930s design clichés. Streamlining is used here only as a symbol of efficiency. The little sucker feet totally undermine the impression of speed. By contrast, the Close Shaver's aerodynamic shape has the functional virtue of fitting the hand.

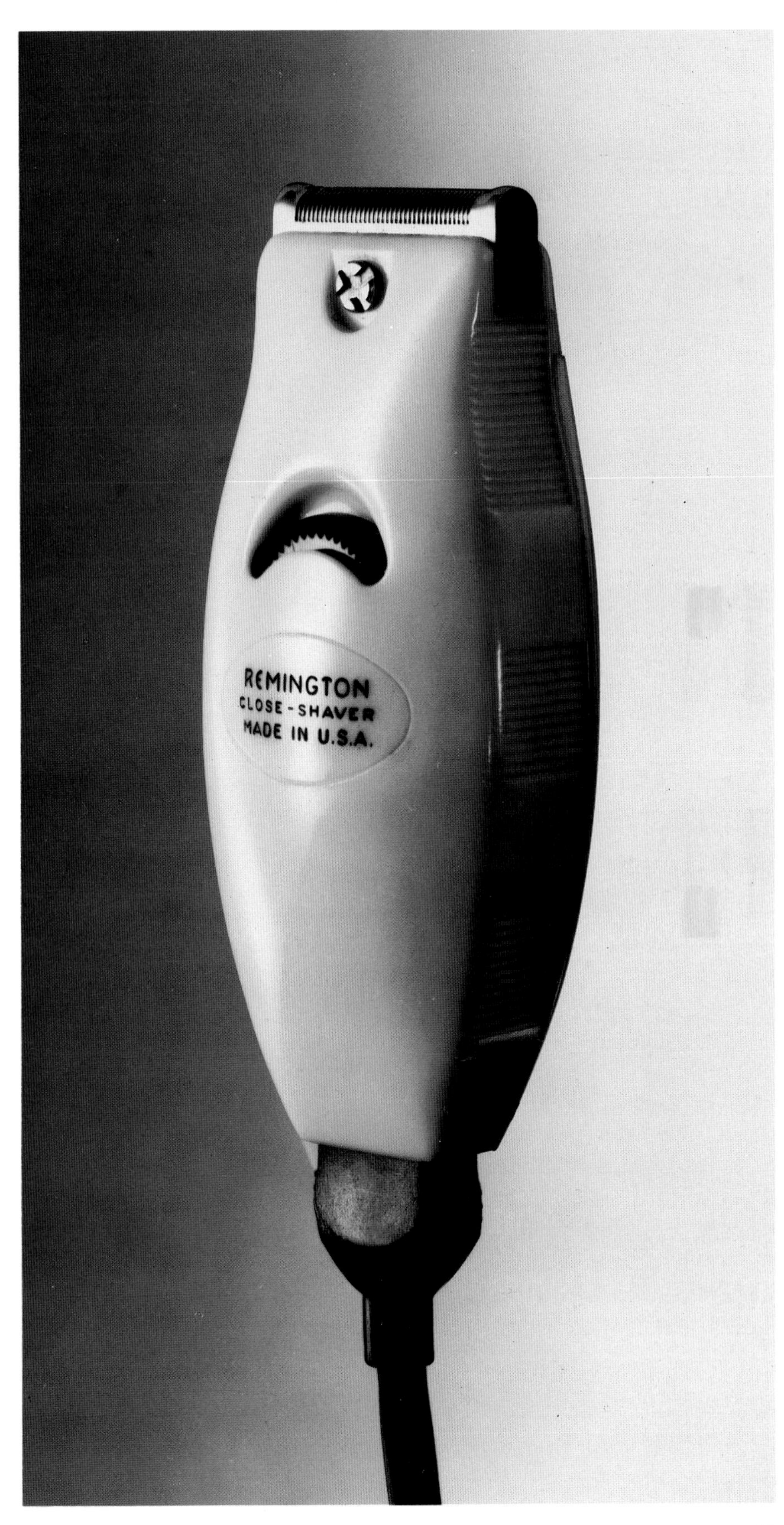

Iron
Black phenolic and chromed steel
USA, 1930s

'Duplicard' duplicating machine
Phenolic
USA, 1935

Dry Shaver
Urea formaldehyde
Manufactured by Remington
USA, 1937

Radio cabinet
Marbled red and orange cast phenolic
Manufactured by Motorola
USA, 1930s

Radio cabinet
Mottled cast phenolic with red handle and knobs
Manufactured by FADA
USA, 1930s

Cigarette box with relief-moulded lid
Urea formaldehyde
Unmarked, probably English, 1940s

4 Economy of effort
The 1940s

War technology and labour-saving for the consumer: plastics as growth industry

An example of compression moulding. The metal mould is elaborately worked with cutting tools to produce a moulding in which polished high relief contrasts dramatically with textured areas.

Acrylic, discovered in the 1930s, came into its own in World War II. Known best by trade names such as Lucite and Perspex, it is a warm, glossy, attractive material with many domestic uses. After the War it was one of the new materials that made transparent furniture possible for the first time. Other new plastics brought see-through films that could be machined into clothes.

Hollywood style bedroom furniture
Acrylic
USA, 1940s

Utility lantern
Tenite cellulose acetate or cellulose acetate butyrate
Manufactured by Tennessee Eastman Corporation
USA, *c.* 1947

Umbrella
Koroseal PVC by B.F. Goodrich Chemical Co.
Manufactured by Halstead & Gravenstein
USA, mid 1940s

Raincoat
Pliofilm transparent rubber hydrochloride film by Goodyear Tire Co.
Manufactured by Richard Boggs & King
USA, mid 1940s

Collapsible hat box
Klearsight film
USA, *c.* 1945

The ease of working acrylic made it possible to use it to make the twisted shapes previously associated with wrought iron and barley-sugar – a thoroughly popular, vernacular style which was liked by the public and scorned by the arbiters of taste. It also lent itself to engraving; as with glass, the worked area shows up as translucent against a clear background.

Handbag
Engraved, clear and patterned acrylic sheet
USA, 1940-50

Handbag
Mottled pearl celluloid sheet with clear acrylic top
USA, 1940-50

Tray, fruit bowl and lamp
Tray and bowl heat shaped from tinted and clear acrylic; lamp base made of glued acrylic sheet and cast phenolic rod with celluloid shade
Probably English, 1940s

As early as 1942 Earl Tupper developed a method of producing thin-walled Polythene containers for which he devised an airtight lid. He also took up the technique of selling direct to the housewife – and indeed of getting the housewife to sell to her friends: and so Tupperware was born. From the few initial bowls and canisters it grew into a whole range, sporting names like 'Millionaire Line', and including a number of designs that have become recognized as classics in plastics.

Charles Eames' chair was equally revolutionary, in its way. The design reflects his wartime background in aeronautical engineering. The one-piece shell, which uses a new material, glass-reinforced polyester (wrongly called fibreglass), has been imitated since to the point of cliché.

Kitchen containers and implements
Polythene
Designed by Earl S. Tupper
Manufactured by Tupperware Plastics Co.
USA, 1946-56

DAR armchair
GRP shell on struts of steel rod
Designed by Charles Eames
Manufactured by Herman Miller Collection
USA, 1948

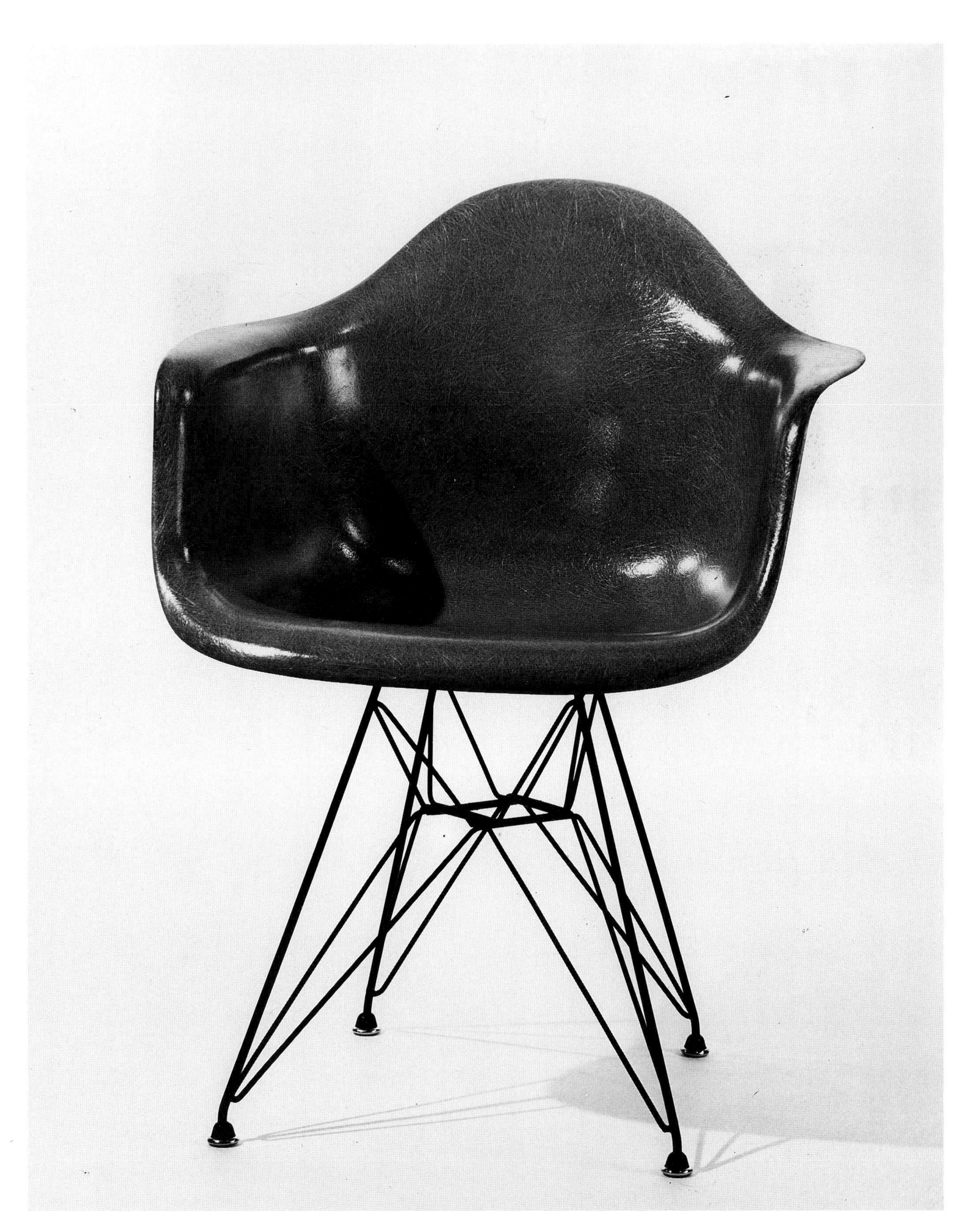

Alongside Polythene, the 1940s brought in PVC (vinyl), a tacky, smelly plastic that suffered in the beginning from instability. But it was strong and waterproof, and it made pocket plastic raincoats possible. The sober, if slightly stretchy, PVC belts belong to the period of austerity and standardized 'Utility' design in Britain.

Tupperware Wonderlier set of bowls
Polythene
Designed by Earl S. Tupper
Manufactured by Tupperware Plastics Co.
USA, 1949

Two belts and cosmetic box
Natural and pearlized PVC belts, polythene box
England, 1940s, 1950s (box)

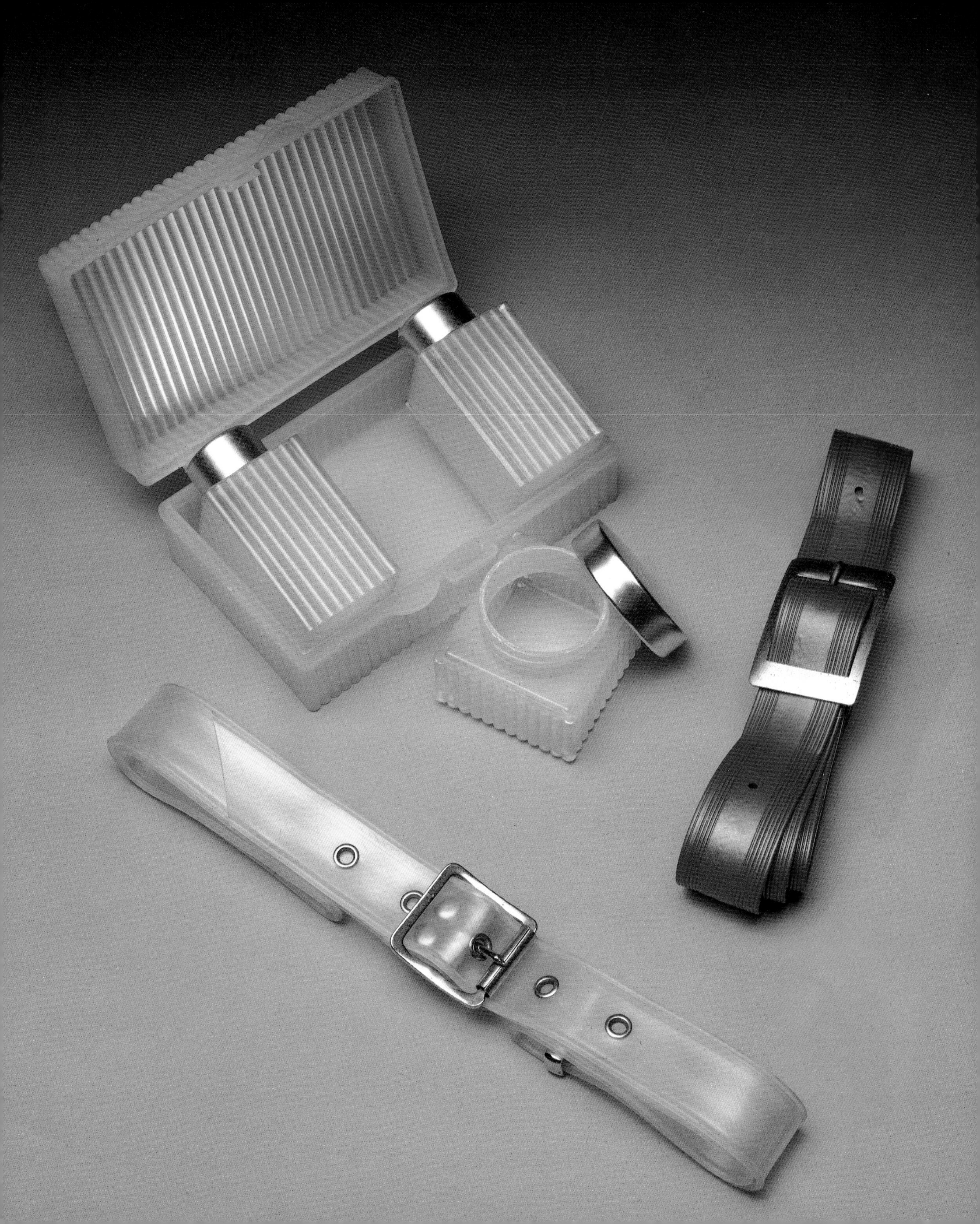

The toys could have been made of celluloid (cellulose nitrate), but its 1940s replacement, cellulose acetate, had the enormous advantage of being nonflammable.

Doll
Tenite cellulose acetate, blow moulded and painted
Manufactured by Tennessee Eastman Corporation
USA, *c.* 1947

Bugle Boy toy trumpets
Tenite cellulose acetate
Manufactured by Tennessee Eastman Corporation
USA, 1940s

British designers in the 1940s, unlike their American counterparts, still favoured dark phenolic. The Bush television, from the early days of television in Europe, could pass for an Art Deco beehive. It now fetches the same high prices as the 1930s designs it resembles, those of the age of Bakelite. By contrast, the small portable radios of the 1940s were rather like handbags, one of them with curious radiator-like fins. The Murphy's ingenious cylindrical control panel leaves no room in the design for a handle, so this is a clutch radio.

Perhaps the most gorgeous plastic objects ever made are the jukeboxes of the 1940s, with their decorative panels in every available colour of cast phenolic and acrylic.

TV12 table television set
Phenolic
Manufactured by Bush Radio
England, 1949

Radio Receiver Type A100
Phenolic
Designed by A.F. Thwaites
Manufactured by Murphy Radio
England, 1946

Radio
Urea formaldehyde
USA, 1940s

Model P147 juke box
Illuminated translucent acrylic
Seeburg Corporation
USA, 1948

SEEBURG

Cellulose acetate plastics were combined with forms of celluloid in both these ingeniously compact designs: the leatherette trim of the radio, and the clear geometry instruments of the Rolinx set, are cellulose nitrate. Rolinx specialized in plastic roll-tops, including some startling cigarette boxes.

Radio
Tenite cellulose acetate or cellulose acetate butyrate by Tennessee Eastman Corporation
Manufactured by Sentinel
USA, *c.* 1947

Rolinx geometry set with Patent Roll-Top Lid Model No. 4
In England the box was made of Bexoid cellulose acetate by B.X. Plastics; the export model was made of cellulose acetate butyrate by Eastman Kodak
Designed by M.H. Robin
Manufactured by Rolinx
England, 1949

BRITISH MAKE ✕ "IMPERIAL MASTER" ✕ PENCILS LTD. LONDON. HB
Rolinx

Three injection-moulded items: elegant design in the cheapest of plastic materials. The streamlined, mollusc-shaped Jumo lamp is adjustable and packs neatly away. The all-plastic mixer, assembled from at least six mouldings, is for delicate tasks like making mayonnaise; the lid holds oil, which drips through a little tube. The cruet set was originally moulded in urea formaldehyde; it is still in use, but already a classic.

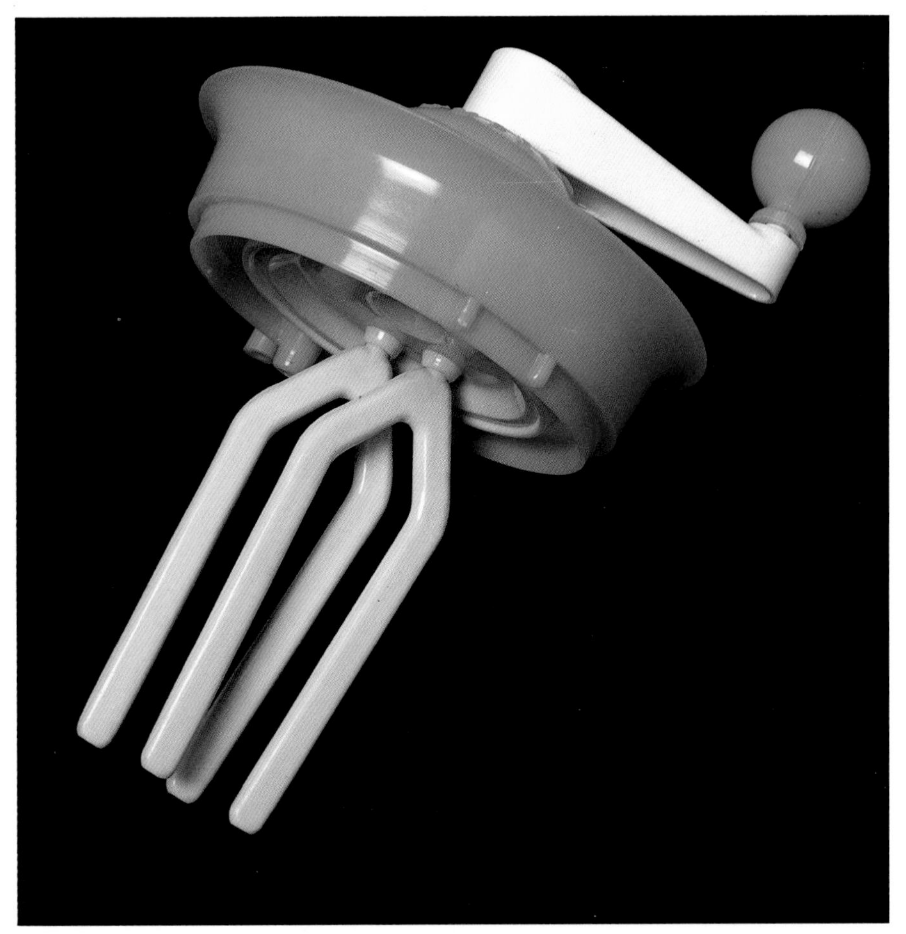

Jumo desk lights
Dark phenolic and white urea formaldehyde
France, 1945

Supermix food mixer
Injection moulded polystyrene
England, 1953

Cruet set
Beetle Melmex melamine
Designed by A.H. Woodfull
Manufactured by BIP Ltd
England, original moulded in urea formaldehyde 1946, melamine early 1950s

The network of Tupperware parties, complete with party games, was started in 1946. It grew so fast that from 1951 it was organized by a special outfit called Tupperware Home Parties.

Tupperware Home Party in Sarasota, Florida, USA, 1958

Shuttlecocks
Injection moulded polythene
England, *c.* 1957

5 The sheerer era
The 1950s

Slick packaging: new discoveries by the minute, as plastics invade every area of life

A single-shot moulding here replaces a complicated mix of feathers, cork and other materials.

Three faces of the 1950s. The elegant little three-legged fruit bowl could not be anything but conscious contemporary design; yet the timeless popular style of the box, with its intricate detail in a lightweight material, was equally typical of the period. The handsome pitcher is visually enhanced by ribs which are also structural features made necessary by the flexibility of Polythene.

Fruit bowl
Polythene on metal 'cherry-stick' legs
England, 1950s

Ornate jewellery casket
Black and pearlized green polystyrene
England, 1950s

Jug
Polythene
England, early 1950s

The Braun food mixer epitomizes the purist design philosophy of the firm; it is still in production. Plastics allow it to be assembled without screws or bolts. The European 1950s also saw the advent of Kartell's first 'designer' bucket, the first bucket to be acquired (other than for domestic use) by the Museum of Modern Art, New York.

KM321 food mixer
Polystyrene housing
Designed by Gerd Alfred Müller
Manufactured by Braun
Germany, 1957

Bucket with lid
Polythene
Designed by Gino Colombini
Manufactured by Kartell-Samco
Italy, 1954

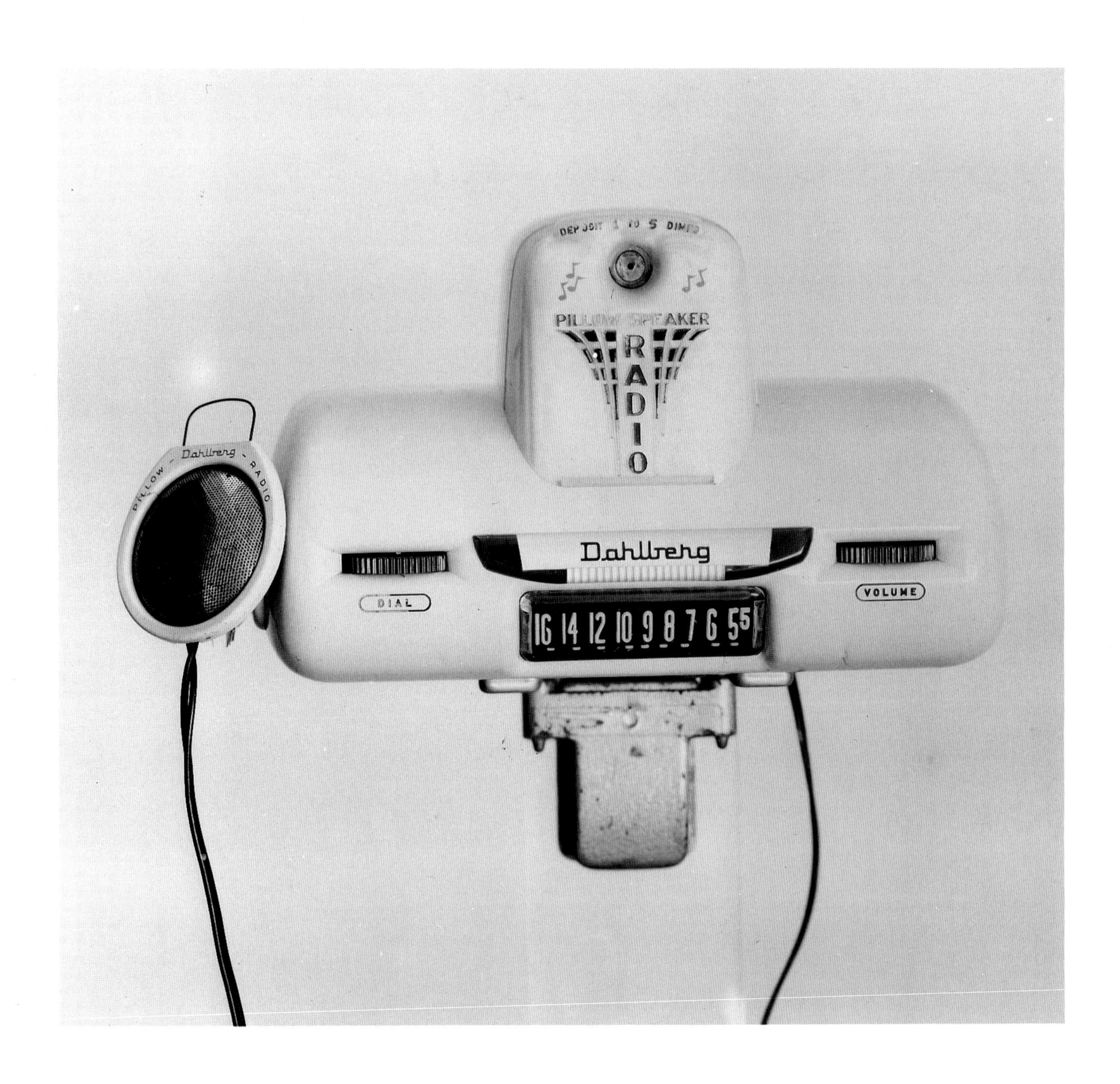

Pillow radio
Urea formaldehyde
Manufactured by Dahlberg
USA, 1950s

By contrast, vernacular design in the USA still tended towards streamlining, as in the coin-in-the-slot radio which has something of the crude commercial form of a shopkeeper's scales.

In the 1950s, stylish streamlined melamine tableware helped to elevate plastics to the dining table. It was much more expensive than ceramics, and many of these indestructible items are still in use. The two-colour effect is achieved by using two mixes in one mould.

The Bourn-Vita cocoa mug started as a two-dimensional image in an advertising campaign: the designer's problem was how to capture the graphic idea in a moulding that would be cheap and simple to produce.

Bourn-Vita Sleeping Beaker
Beetle urea formaldehyde, with nightcap in blue polythene and bobble in red cellulose acetate
England, 1951

Part of the Melaware range of tableware
Melamine, two-colour mouldings
Designed by A.H. Woodfull
Manufactured by Ranton & Co, distributed by Melaware
England, 1959

Part of the Fiesta range of tableware
Melamine
Designed by Ronald E. Brookes
Manufactured by Brookes & Adams
England, 1961

DAF chair
GRP shell on chromium-plated metal legs
Designed by George Nelson
Manufactured by Herman Miller Collection
USA, 1956

View of the Exhibition of Science, part of the Festival of Britain, London 1951
600-foot-long light fitting based on the atomic structure of carbon, the element common to most plastics
Designed by Brian Peake

Gaby Schreiber pioneered a process of coloured decoration on Runcolite plastics tableware which, like underglaze painting on china, would resist rubbing.

The ungainly tripod lighter is a period piece, typical of much top-heavy 1950s design, and probably a lot more popular with today's collectors than it ever was with smokers.

Three-legged table lighter
Urea formaldehyde
Manufactured by Beney
England, 1955

Festival of Britain souvenir cocktail tray
Compression moulded ivory urea formaldehyde
Designed by Gaby Schreiber, with centre emblem designed by Abram Games
Manufactured by Runcolite
England, 1951

Tray
Compression moulded urea formaldehyde
Designed by Gaby Schreiber, with farmyard animals by Pauline Baynes
Manufactured by Runcolite
England, 1951

Stellar fantasy: the camera, which takes four pairs of 3-D photographs, definitely suggests the more frivolous end of the science-fiction market. The early transistor radio has a curious decor like a radar-screen or a cloud-chamber.

The Vogue label made picture discs for only a few years in the 1950s. This one includes an other-worldly landscape with a Surrealist melting sundial.

3-D camera
Multichrome urea formaldehyde
Manufactured by Coronet
England, 1950s

Cordless transistor radio
Urea formaldehyde case with decorative laminate panel
Manufactured by General Electric
USA, 1950s

Picture record of 'This Is Always'
PVC
Manufactured by Vogue
USA, *c.* 1953

THIS IS ALWAYS
MACK GORDON—HARRY WARREN
JOAN EDWARDS
ACCOMPANIED BY THE VOGUE RECORDING ORCHESTRA
Vogue
THE
PICTURE RECORD
A TOM SAFFADY PRODUCT
LICENSED FOR NON-COMMERCIAL USE ON HOME PHONOGRAPHS ONLY.
R767

Resealable bottle closures became an important use of Polythene – even if these early efforts look rather unglamorous. Meanwhile, cosmetics firms continued to use plastic for elegance.

Orchis perfume bottle and packaging
Urea formaldehyde and glass
Designed by Reco Capey
Manufactured by Yardley
England, 1954

Droppers for eau-de-Cologne, perfume and brilliantine
Polythene
France, *c.* 1950

Queen Elizabeth II coronation badges
Printed art paper clamped to tinplate base with celluloid sheet
Designed by John Bruce
Manufactured by Alfred Roden
England, 1952

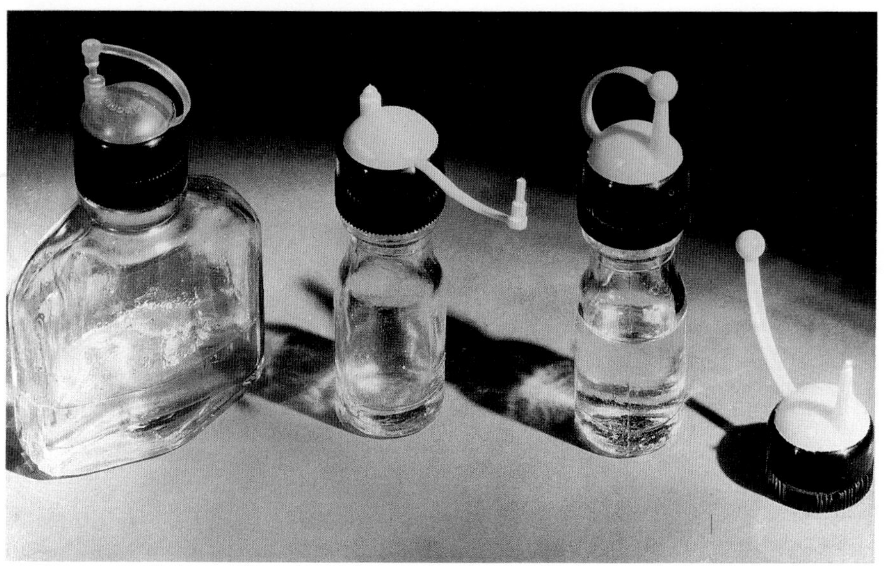

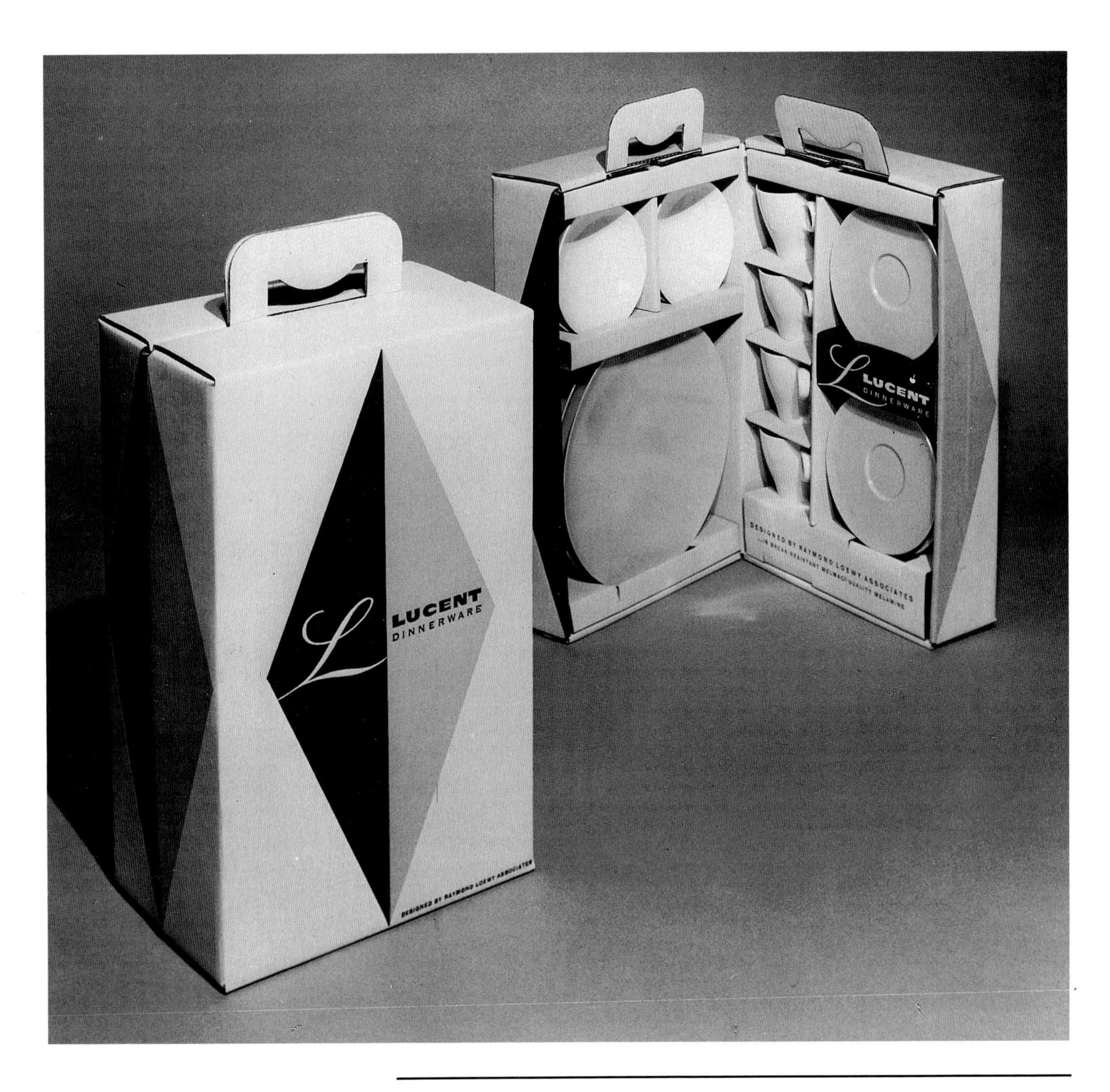

The stylish Lucent dinnerware pack, by the leading US industrial designer Raymond Loewy, won an award for the double feat of bringing 'new dignity and a larger market to this type of product'.

Packaged dinnerware set
Melamine
Packaging designed by Raymond Loewy Associates
Dinnerware manufactured by Lucent
USA, 1956

TTK TR-63 transistor radio
Manufactured by Sony
Japan, 1957

Midge radio
Phenolic
Manufactured by Northern Electric
USA, *c.* 1950

2+7 telephone
Cellulose acetate housing
Designed by Marcello Nizzoli
Manufactured by SAFNAT
Italy, 1958

Aerodynamics and compact form: the Midge and the 2+7 make compactness almost a mannerism. By contrast, the firm contours of the first pocket transistor radio mark the beginning of a long line of classic Japanese miniatures.

MIDGE
Northern Electric

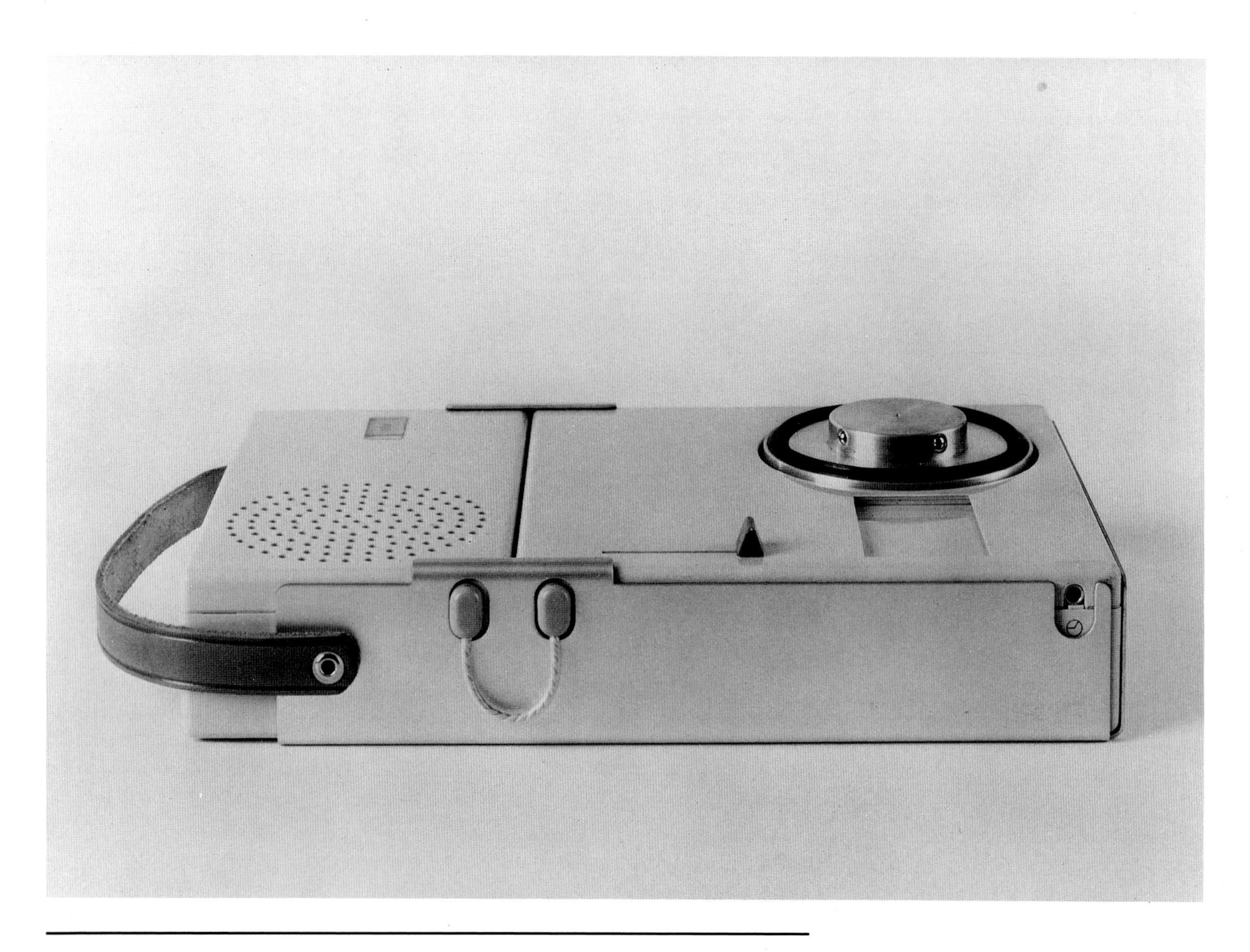

Braun's battery-operated miniature record player won a design award at 'Interplas', London 1961. The sliding control brings up the stylus, which plays the record from below; unwary users thus tend to get the flip side by mistake.

Combined transistor radio and record player
Two-part polystyrene case
Designed by Dieter Rams
Manufactured by Braun
Germany, 1959

Superonda (Superwave) sofa
PVC upholstered polyurethane foam
Designed by Archizoom
Manufactured by Poltronova
Italy, 1968

6 Good-time gloss
The 1960s

From fun furniture to cast acrylic candy:
inflatables, disposables and the wet look

A 1960s 'dolly-bird' tries to look comfortable on clammy PVC. The blocks can be placed in different arrangements.

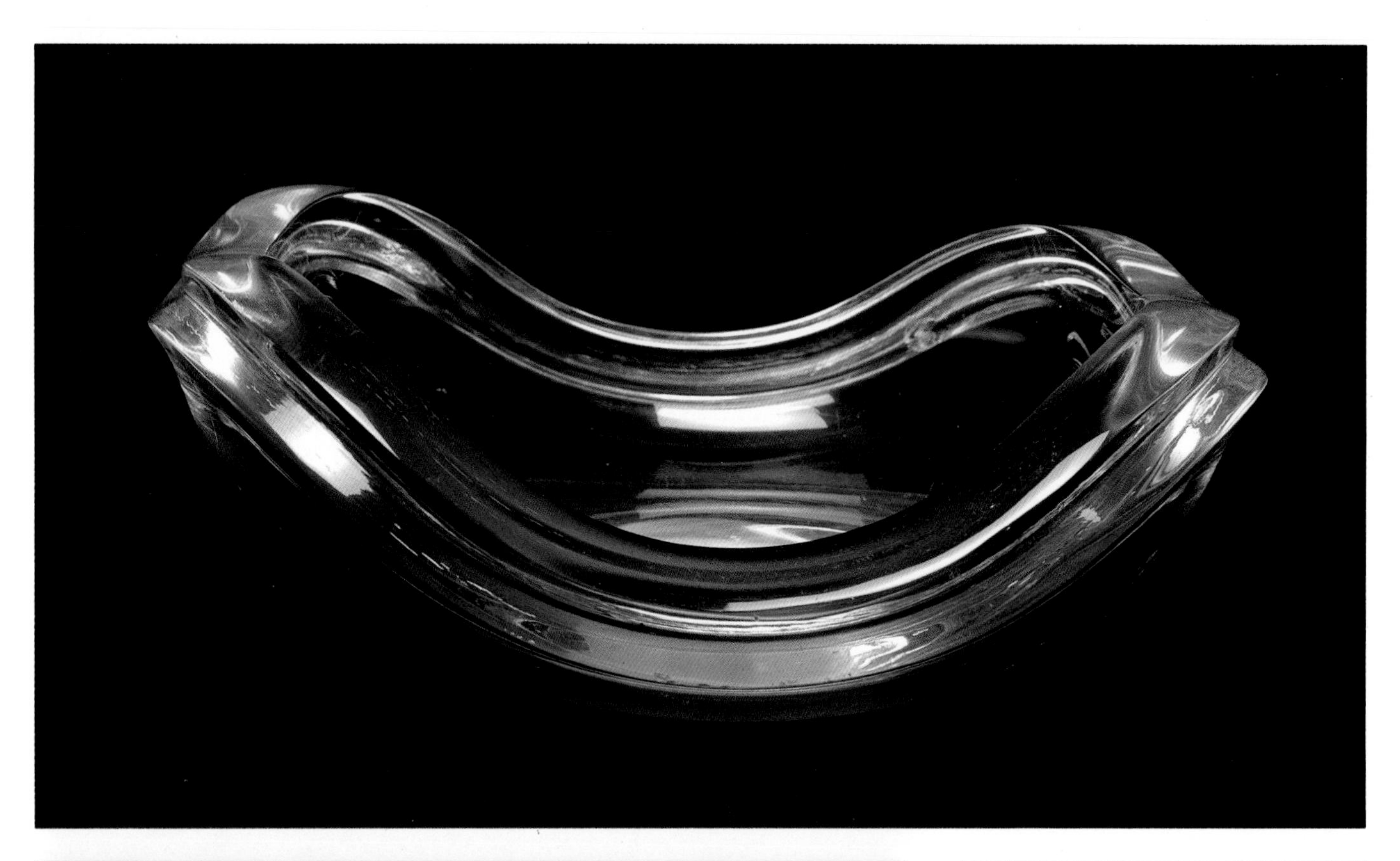

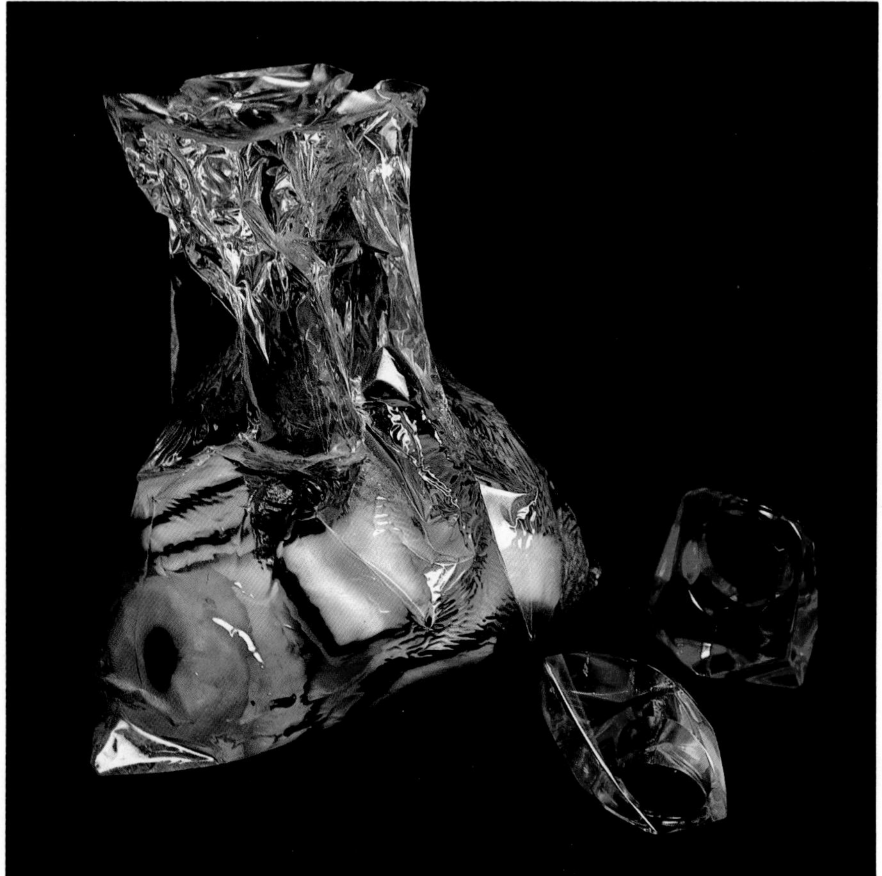

One of the expensive plastics, acrylic can be cast cold – even in a cellophane candy bag embedded in gravel. Cast acrylic can also be machined into all sorts of sculptural shapes.

Astrolite fruit bowl
Solid machined acrylic
Manufactured by Ritts Co.
USA, 1960s

Liquorice Allsorts sweetbag
Real sweets embedded in acrylic
Designed by Paul Clark
Manufactured by Objex
England, made from early 1970s until 1980

Two rings in 1960s style
Clear and coloured laminated acrylic
Taiwan, 1983

Small table
Heat shaped acrylic
1963

The sharp, shiny wet-look image of the 1960s was created entirely by plastics. By the time a man first set foot on the moon in 1969, space-age materials, and concepts of space, had changed the face of design. Transparent materials such as acrylic and PVC were perfect for giving the illusion of floating in space on non-existent furniture, for revealing the inner workings of clocks, television sets and even wardrobes, and of course for exhibiting the body inside see-through clothes.

Section of a wardrobe
Acrylic
Designed by Leonardo Fiori
Manufactured by Zanotta Poltrone
Italy, 1968

Magazine rack
Clear and bronze acrylic sheet fixed with chromed connectors
Designed by Kenneth Brozen
Manufactured by Robinson-Lewis-Rubin
USA, 1968

Dining chair
Clear acrylic sheet with chamois leather upholstery
Designed by Casati-Ponzio
Manufactured by Comfort
Italy, 1968

Low easy chair
Clear acrylic sheet with chamois leather upholstery
Designed by Casati-Ponzio
Manufactured by Comfort
Italy, 1968

Sunglasses have long been a fashion item, constantly changing with every whim. These represent both 1960s high fashion (the space-age black wraparounds) and swinging assymmetry. In the midst of all this, 1960s Tupperware cups (sold with caps for cold storage) contrive to maintain their image of cut-price gracious living.

Tupperware Parfait Set
Frosted polythene cup with snap-fit base
Manufactured by Dart Industries
USA, 1968

1960s sunglasses
Clear acrylic asymmetrical sunglasses made by Pierre Cardin, France; pearlized cellulose acetate Flash Gordon style sunglasses made in Italy; three pairs of sunglasses in black polystyrene and cellulose acetate made in France and Korea
1960s

Fun and freedom: Blow was the first inflatable chair on the market in Europe. Fun furniture was never made to last, and it is hard to find now. It graced the same 1960s interiors in which Boalum liberated lighting from its fixed relationship to four walls.

Blow chairs
Inflatable PVC
Designed by Scolari, Lomazzi, D'Urbino and De Pas
Manufactured by Zanotta Poltrone
Italy, 1967

Seating unit
Inflatable transparent PVC
Manufactured by Mobilier International
France, 1968

Boalum snake light
Flexible PVC hose with embedded spiral, 200 cm long, 6 cm dia.
Designed by Livio Castiglioni and Gianfranco Frattini
Manufactured by Artemide
Italy, 1969

Thanks to plastics, 1960s chair design took off into strong colours and completely new forms. The principle linking all these designs was that the chair should give the impression of being a single organic shape, even if it was not. On the right is the first one-piece moulded chair.

Chair 582
Rubber sheet over tubular metal frame, upholstered with jersey-covered latex foam
Designed by Pierre Paulin
Manufactured by Artifort
Holland, 1965

Fiorenza chaise longue
Thermo-formed ABS
Designed by Motomi Kawakami
Manufactured by Bazzani
Italy, 1968

Child's stacking chair
High density polythene
Designed by Marco Zanuso and Richard Sapper for the Municipality of Milan
Manufactured by Kartell
Italy, 1961

Flowing form: 1960s produced the first one-piece stacking chair. It looks hard, but is comfortable and springy to sit on. The other four forms illustrate some of the huge variety of plastic materials that were available, and the ways in which designers enjoyed their new freedom.

Stacking chair
Originally moulded in GRP
Designed by Verner Panton
Manufactured by Herman Miller Collection
Germany, 1960

PH 6 easy chair
Heat-shaped acrylic sheet fixed to chromed frame
Designed and manufactured by Peter Hoyte
England, 1968

CL9 Ribbon chair
Continuous strip of GRP cantilevered from a tubular steel frame
Designed by Cesare Leonardi and Franca Stagi
Manufactured by Elco
Italy, 1969

'Nike' chaise longue
Vacuum formed ABS, base filled with rigid polyurethane foam, on enamelled steel base
Designed by Richard Neagle and E. Szego
Manufactured by Sormani
Italy, 1968

Chair 4860
Originally moulded in ABS, now in polypropylene
Designed by Joe Colombo
Manufactured by Kartell
Italy, 1965

Irreverent machines: the construction kit makes a robot which stamps its feet, waves its arms, spins and flashes its eyes. The clocks are definitely not for people in a hurry.

Three clocks
Dials of PVC on metal stands
Designed by Paul Clark
Manufactured by Perspective Designs
England, (centre) *c.* 1966, (others) *c.* 1971

Children's construction kit
Nylon
Designed by Artur Fischer
Manufactured by Fischer Technik
Germany, *c.* 1970

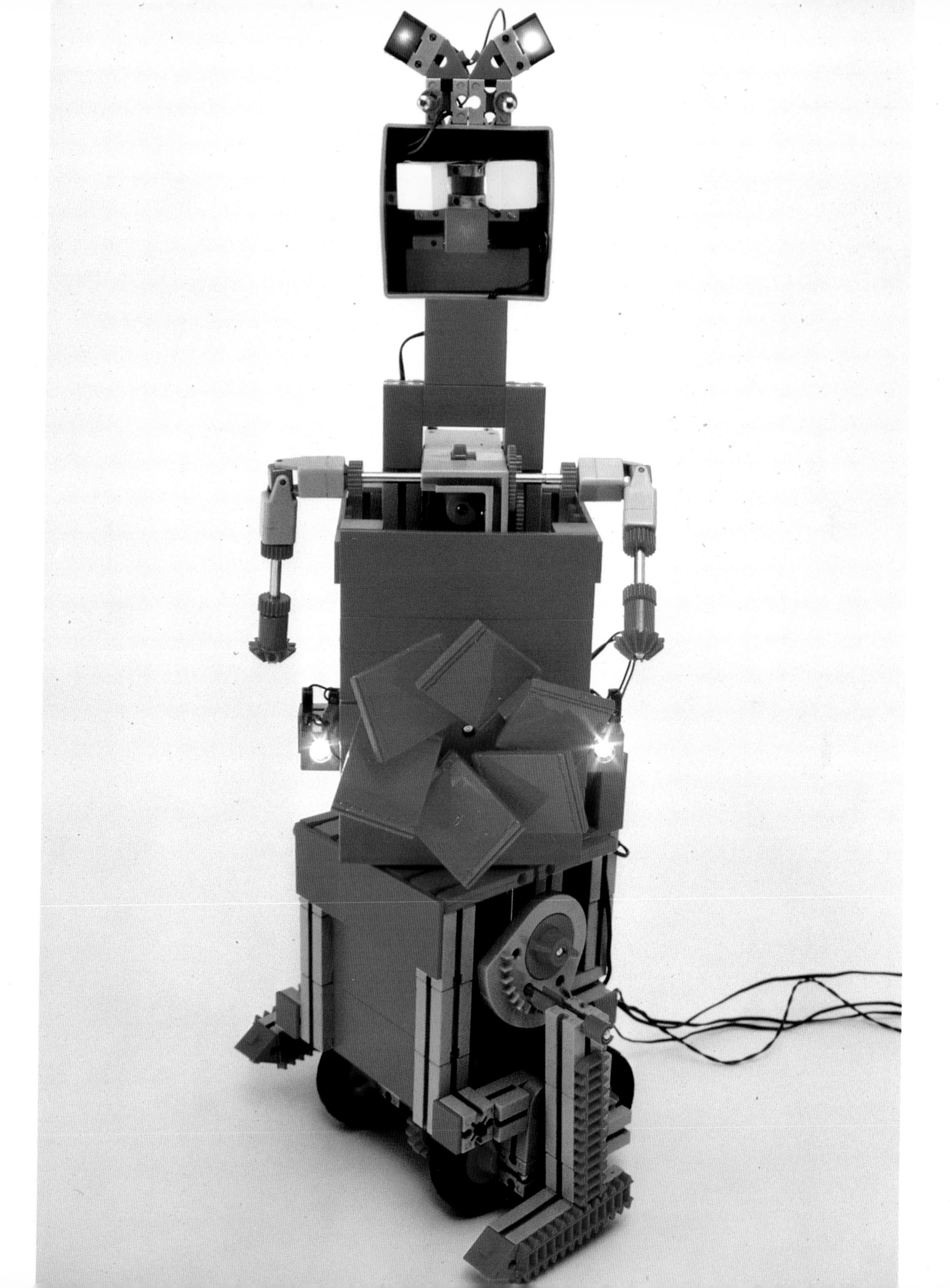

Seating units and hat stand
Upholstered and self-skinned polyurethane foam
Designed by Gilardi, Ceretti, Derossi and Rosso
Manufactured by Gufram
Italy, 1969-70

Sassi (Stones)
Self-skinned polyurethane foam
Designed by Piero Gilardi
Manufactured by Gufram
Italy, 1967

Polyurethane foam, which can produce its own skin in a hot mould, will make sculptures that are also furniture: impossible, inhospitable shapes that nevertheless yield when you sit on them. In some, a skin holds a soft outer layer and a firmer core.

'Executive' pocket calculator
ABS
Manufactured by Sinclair Radionics
England, 1972

7 A transitory presence
The 1970s

An end to abundance: cynical sophistication and the rise of structural foams

Clive Sinclair, computer whizz-kid of the 1970s and 1980s, designed his first calculator – the world's thinnest and lightest, and the first real pocket calculator – in 1972. It was snapped up for the collection of the Museum of Modern Art in New York.

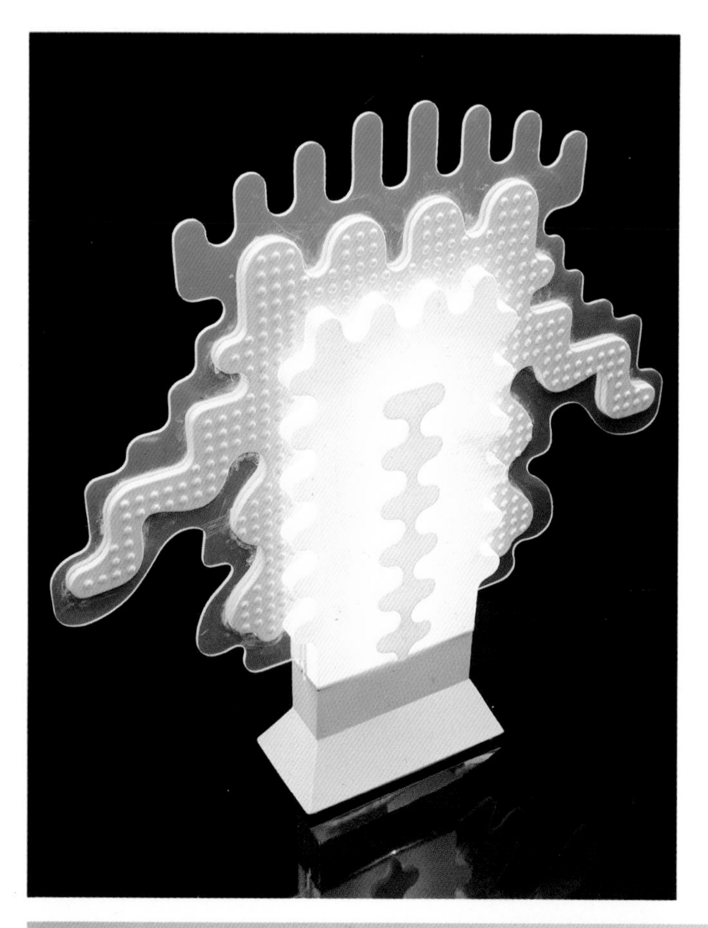

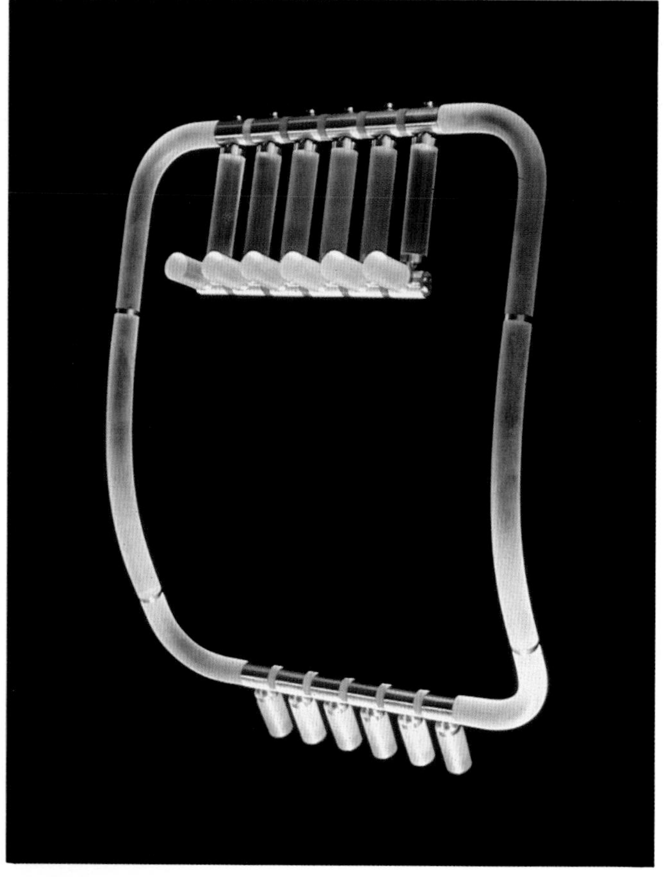

The unpredictable 1970s: at first sight it is hard to guess what these shiny shapes are.

Light
Vacuum formed ABS and acrylic
Designed by George J. Sowden
Manufactured by Planula in limited edition of 35
Italy, 1972

Radio
Components cast in acrylic
Designed by David Watkins
Manufactured by Clarity Plastics
England, 1973

Body-form necklace
Sandblasted acrylic rod and silver
Designed by David Watkins
England, 1974

Hairslide
Various plastics including printed acetate sheet, polypropylene sheet, cellulose acetate and a knitting needle
Designed by Ally Capellino
England, 1979

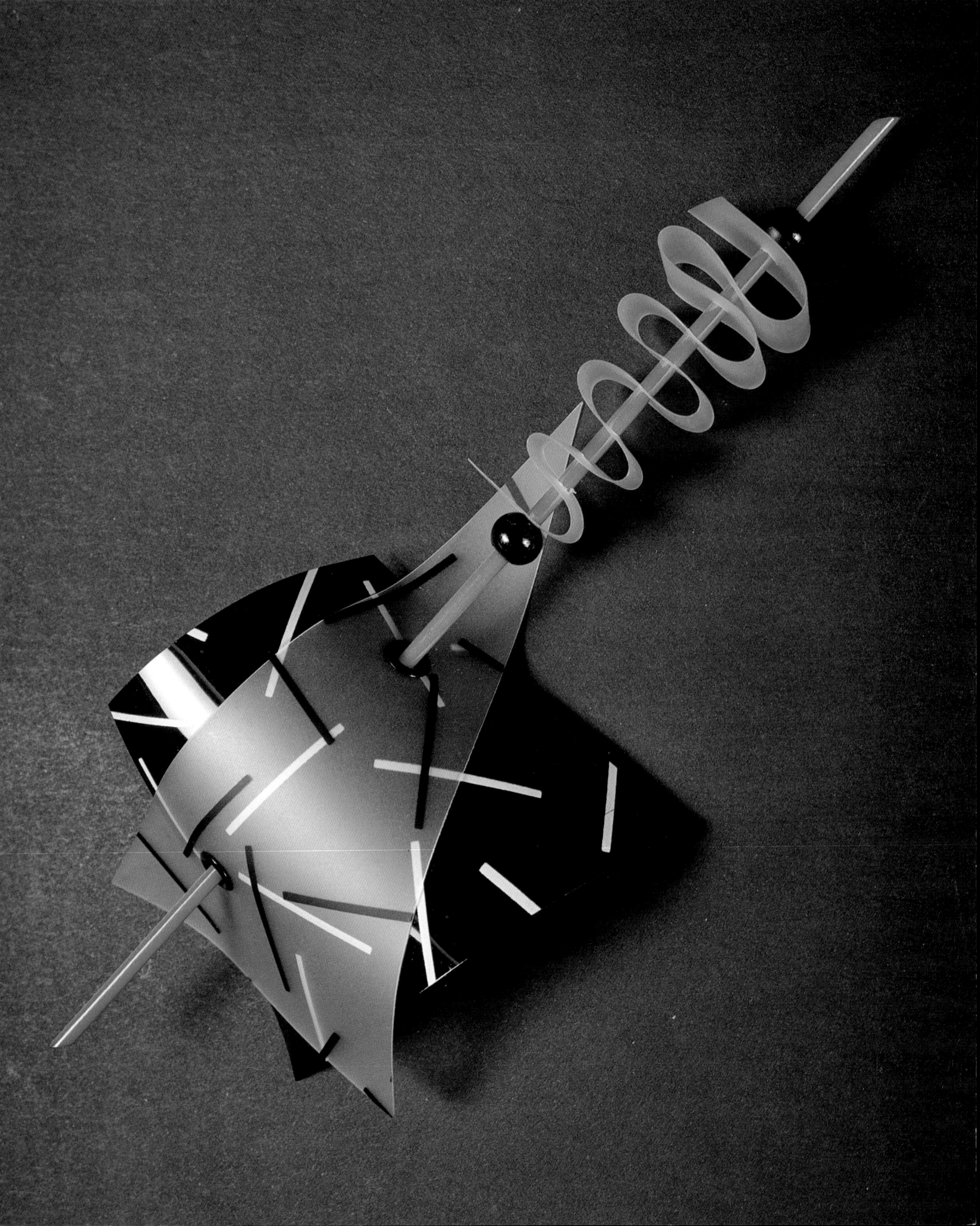

Bendit chair
Sheet of polythene, or polypropylene, or PVC, or cellulose acetate butyrate, screwed to metal frame
Designed by Anthony Hopper
England, 1970

Semi-disposable cutlery
Styrene-acrylonitrile
Designed by David Harman Powell
Manufactured by Ekco Plastics
England, 1970

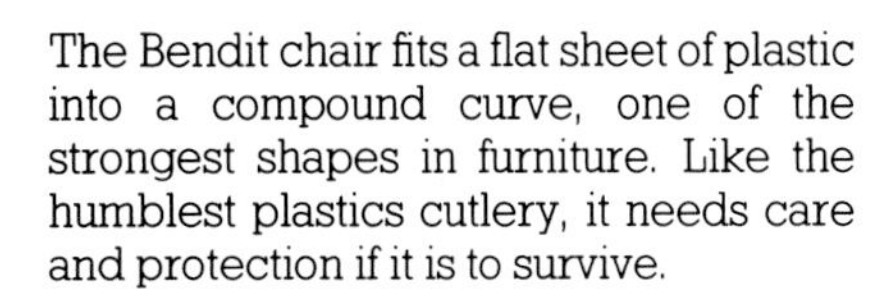
The Bendit chair fits a flat sheet of plastic into a compound curve, one of the strongest shapes in furniture. Like the humblest plastics cutlery, it needs care and protection if it is to survive.

The clocks that use colours and abstract shapes to represent hours and minutes are reminiscent of Bauhaus exercises; and the cut-out foam furniture is more like graphic design than furniture design. With its 1960s contrast of hard form and soft material, it is primarily a shape to play around with.

Wall clocks
Sheet acrylic
Designed by Anthony Gemmill with dials by Paul Clark
Manufactured by Acrylic Products
England, 1970

Travel toothbrushes
Injection moulded dyed nylon
Designed by Jenny Albrecht
Manufactured by Buch & Deichmann
Denmark, *c.* 1978

Spanner and Nut seat/table unit
PVC dipped polyurethane foam
Designed by Rupert Oliver
England, 1971

XA2 camera
ABS
Manufactured by Olympus
Japan, 1979

Countdown CD1 electronic digital clock
ABS with sprayed paint finish
Designed by John Ryan
Manufactured by the House of Carmen
England, 1975

Teneride armchair
Cylinder of self-skinned polyurethane foam mounted on rotating GRP base
Designed by Mario Bellini
Italy, *c.* 1970

The 'Black Box' philosophy runs through much of the design of the late 1970s and early 1980s. It subjugates both material and function to preconceived form: anything can be put in a Black Box, and the result is virtually guaranteed to look efficient. The black foam swivel chair, by contrast, a prototype that never went into production, explores a new and unprecedented form.

Incredible Inedibles: replica food in plastics came from Japan where it was developed to replace wax food offerings at festivals. It has its uses in training aircraft cabin crews and for everlasting window displays.

Synthetic biscuits and bread
PVC painted with plastisol; the bread has a filler of rigid polyurethane foam
Designed by Karen Ringrose
Manufactured by Replica Food
England, from 1977, still in production

Fruit bowl
Plastic laminate and painted steel
Designed by George J. Sowden for 'Objects for the Electronic Age'
Manufactured by ARC 74
Italy, 1983

8 Function and beyond
The 1980s

Solid engineering and ornamental riot: new metaphors and utopias for the plastic future

Far from imitating traditional materials, the designers of 'Objects for the Electronic Age' use the potential of sheet plastic and metal to challenge the way we expect things to look.

Rod, hose, sheet: modern plastics jewellery uses standard pre-formed materials as a starting-point for fantasy.

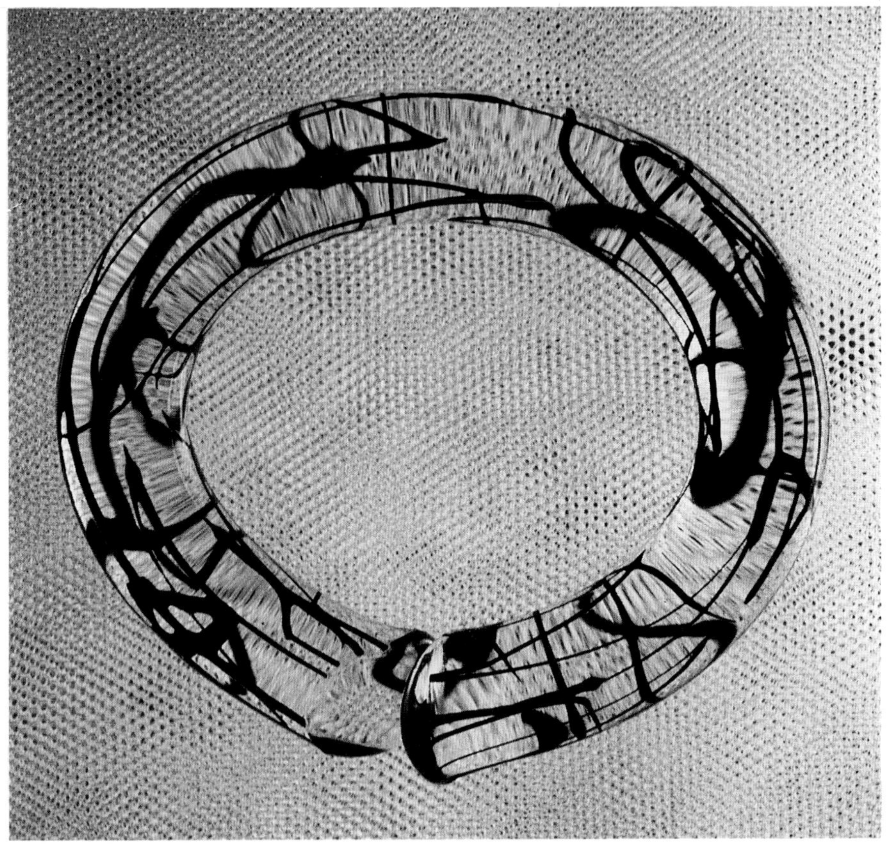

Bangle
Clear acrylic rod painted with acrylic screening ink
Designed by Marlene McKibbin
England, 1983

Bracelet
Corrugated rigid PVC
Designed by Louise Slater
England, 1982-83

Earrings and haircomb
Scored and folded rigid PVC with
cellulose acetate comb
Designed by Louise Slater
England, 1982-83

Right
Earrings
Dyed nylon wire
Designed by Alison Baxter
England, 1981

Left
Bracelet
Cut and polished ebonite inlaid with polyester resin
Designed by Jane Kennard
England, 1982

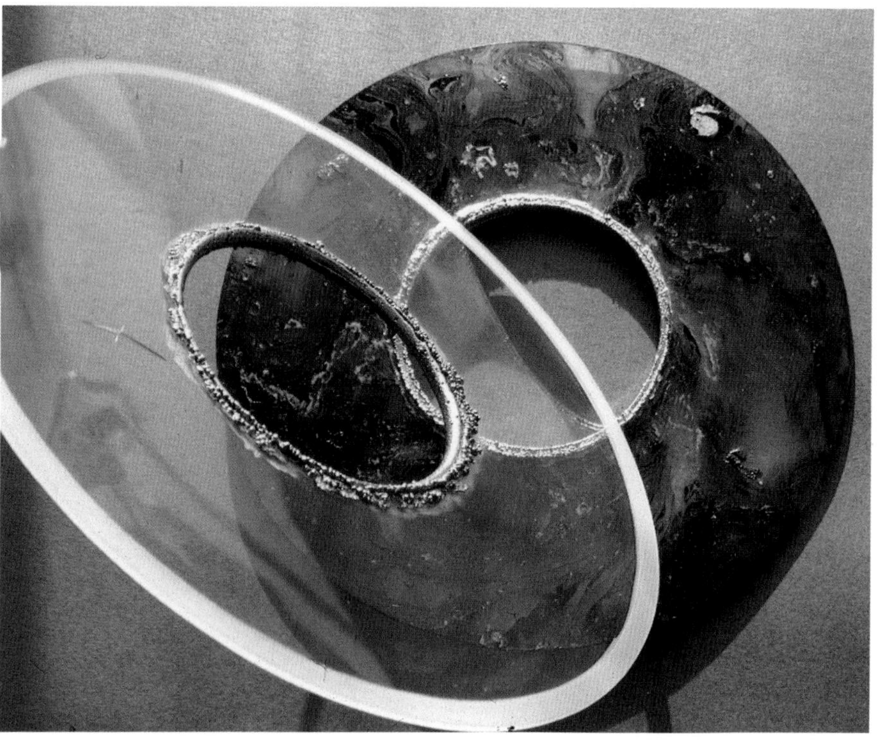

Bangles
Acrylic with electroplated rims
Designed by Bo Davis
England, 1981

Simulated bamboo
PVC
Manufactured by Plastirama
Mexico, 1982

Jane Kennard rescues off-cuts of ebonite (now a rare material) for her inlaid jewellery; Bo Davis experiments with surface treatments of acrylic.

The Transformer mattress/chair demands the participation of every user. The positive body image (background) is entirely non-practical: the negative impression from one Transformer was used as a mould to shape another.

Transformer seat
PVC envelope filled with expanded polystyrene beads, which form a rigid shape when the air is sucked out
Designed by Ron Arad
Manufactured by One Off
England, 1983

Hole in the Wall mirror
Acrylic and plaster
Designed by Lee Curtis and Tom Lynham
England, 1983

Brutalist expressionism with an industrial finish gives the Dalila chairs a timeless if unsympathetic quality.

The jacket light, a more attractive tour de force, is vacuum formed from acrylic sheet round a solid mould.

Dalila chair
Rigid polyurethane with epoxy resin finish
Designed by Gaetano Pesce
Manufactured by Cassina
Italy, 1980

Dalila armchair
Rigid polyurethane with epoxy resin finish
Designed by Gaetano Pesce
Manufactured by Cassina
Italy, 1980

Jacket light
Thermoformed translucent acrylic
Designed by Jacques Vojnovi
France, *c.* 1982

Royal chaise longue
Designed by Nathalie du Pasquier, red/green fabric designed by George J. Sowden
Manufactured by Memphis
Italy, 1983

Table 4300
Top rigid expanded polyurethane; cones ABS; legs polypropylene
Designed by Anna Castelli Ferrieri
Manufactured by Kartell
Italy, early 1983

Chair 4875
Polypropylene
Designed by Carlo Bartoli
Manufactured by Kartell
Italy, 1979

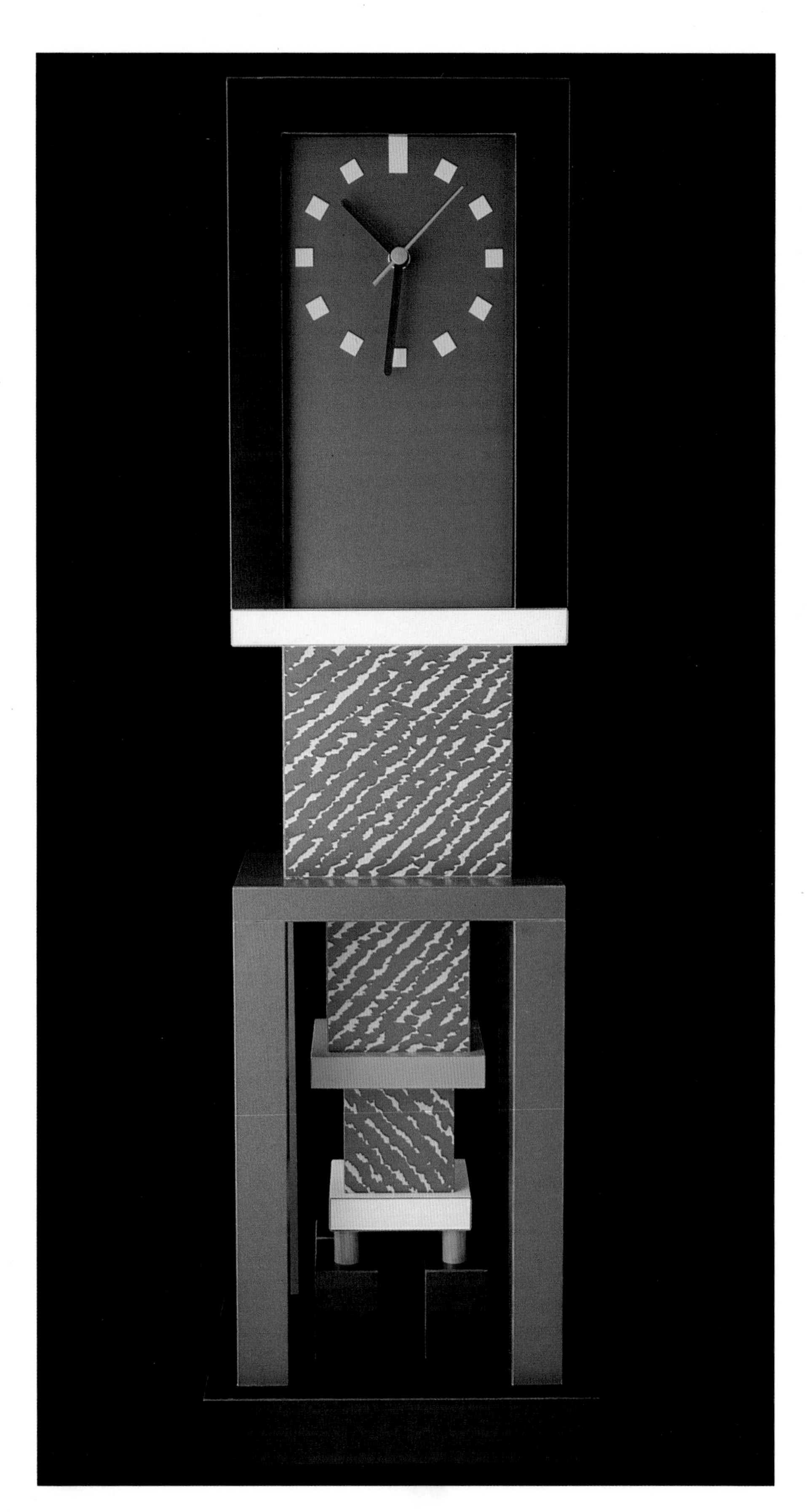

Craquelé fruit bowl
Plastic laminate with nickel-plated steel and painted wooden feet
Designed by Nathalie du Pasquier for 'Objects for the Electronic Age'
Manufactured by ARC 74
Italy, 1983

Metropole clock
Wood with decorative laminate
Designed by George J. Sowden, laminate designed by Christophe Radl
Manufactured by Memphis
Italy, 1982

The impressive 4300 table took a year to design; it has three materials, chosen for their different properties and signalled by differences in colour.

Memphis turned the design world upside down with perverse and impractical forms in their first show in 1981; only a year later, their designs were less outrageous, and there is even a hint of Mackintosh in the clock.

The Mirizzo sofa looks as if it has just landed from outer space; the design and use of materials make a point of its irrationality and hollowness. The knock-down Kartell table is solid ABS, a pointer to the restrained use of plastics in the future, based on careful engineering: the parts slot together without a single metal fixing.

Mirizzo sofa
Marbled plastic laminate with upholstered foam on chromed metal frame
Designed by Mancini
Manufactured by Kaleidos
USA, 1981

Table 4310
Top ABS painted with polyurethane; pedestal and interlocking feet sprayed ABS
Designed by Anna Castelli Ferrieri
Manufactured by Kartell
Italy, 1983

Necklaces, earrings and bracelets
Acrylic rod and cast polyester with routed decoration filled with coloured polyester resin
Designed by Pat Thornton
Manufactured by Cicada
England, 1983

Tape shape eraser
Tinted PVC
Manufactured by Kutsuwa
Japan, 1982

Push Pen ballpoint
Square sided tube operated by compressing rubber bellows
Manufactured by Euroway
Italy, 1982

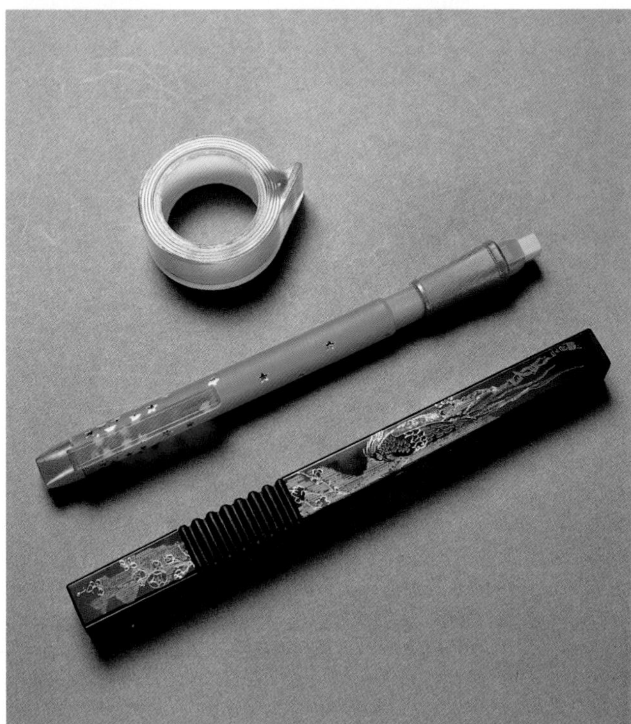

Radio
Envelope of welded PVC
Designed by Daniel Weil
Manufactured by Parenthesis
England, 1981

Radio
Envelopes of welded PVC
Designed by Daniel Weil
England, 1983

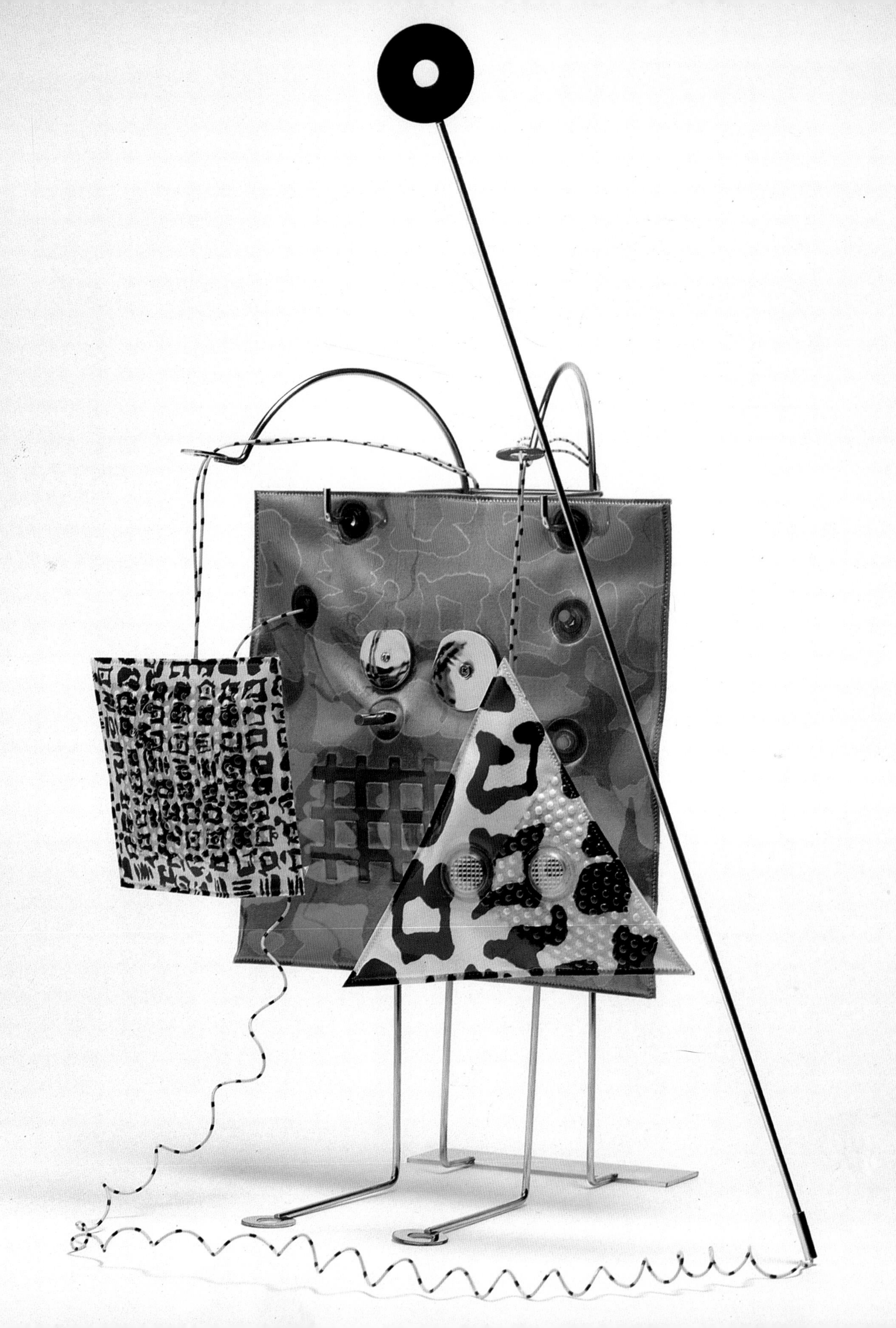

BRAUN

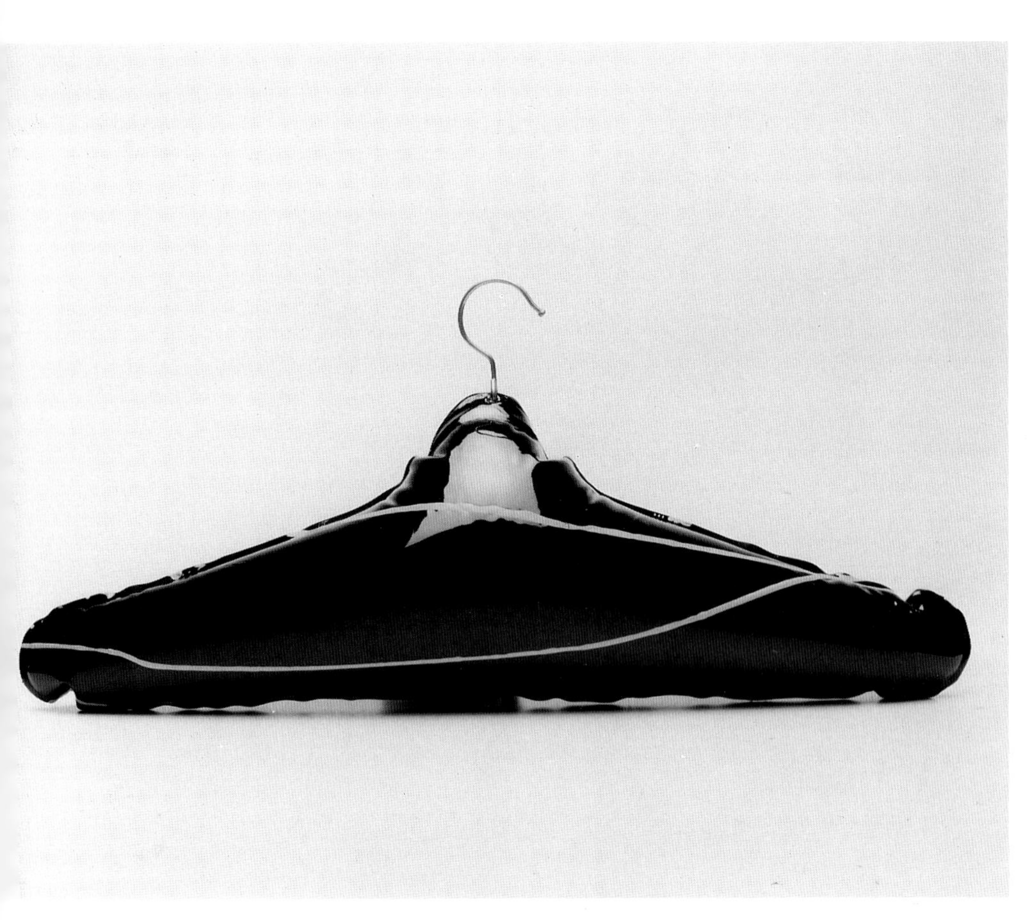

Function and the designer's eye. 'The best design for me is the least design, modest and unimposing, neutral and balanced': the decorative pattern on Dieter Rams' shaver is in fact a functional non-slip polyurethane finish.

Micron Plus shaver
ABS embossed with polyurethane
Designed by Dieter Rams
Manufactured by Braun
Germany, 1980

Inflatable coathanger
PVC film decorated with acrylic paint
Designed by Helen Gripaios
England, 1980

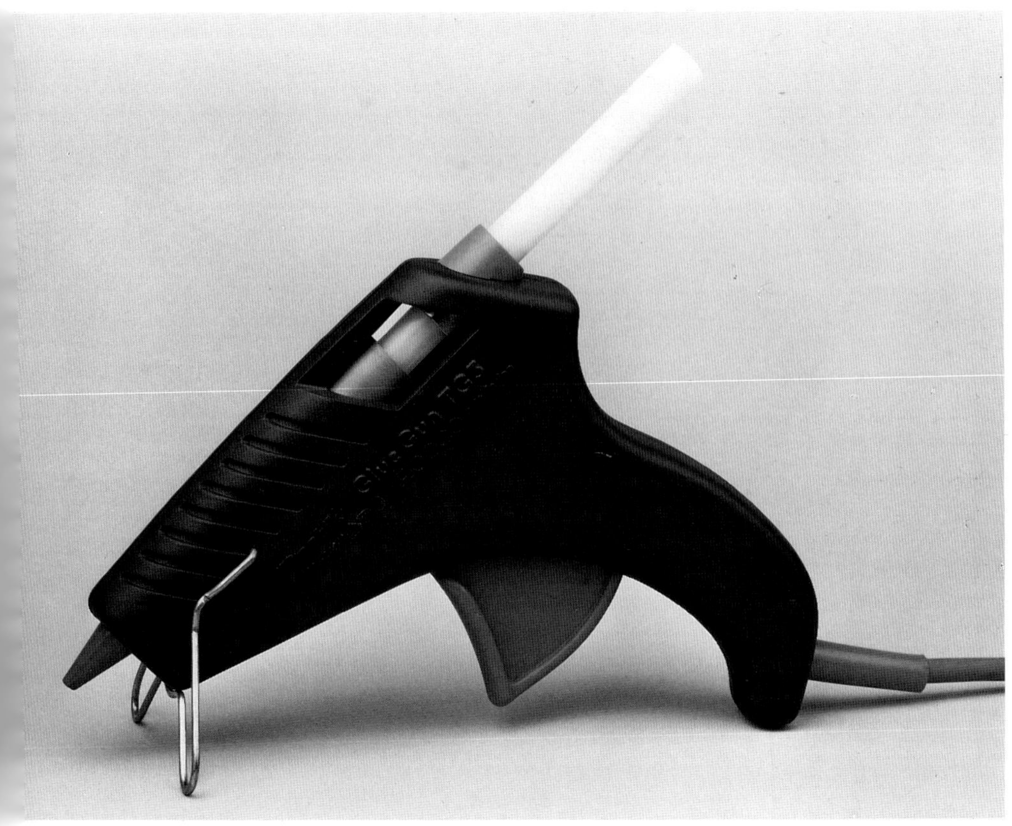

Most glue guns look industrial and rather clumsy; the Bostik design team have made a conscious effort to make theirs easier and more pleasant to use – and to look at. Gripaios' treatment of the inflatable hanger – not a new invention – is frankly decorative.

Trigger glue gun TG3
Advanced reinforced polyester
Designed by Bostik, Germany
Manufactured by Bostik
England, 1983

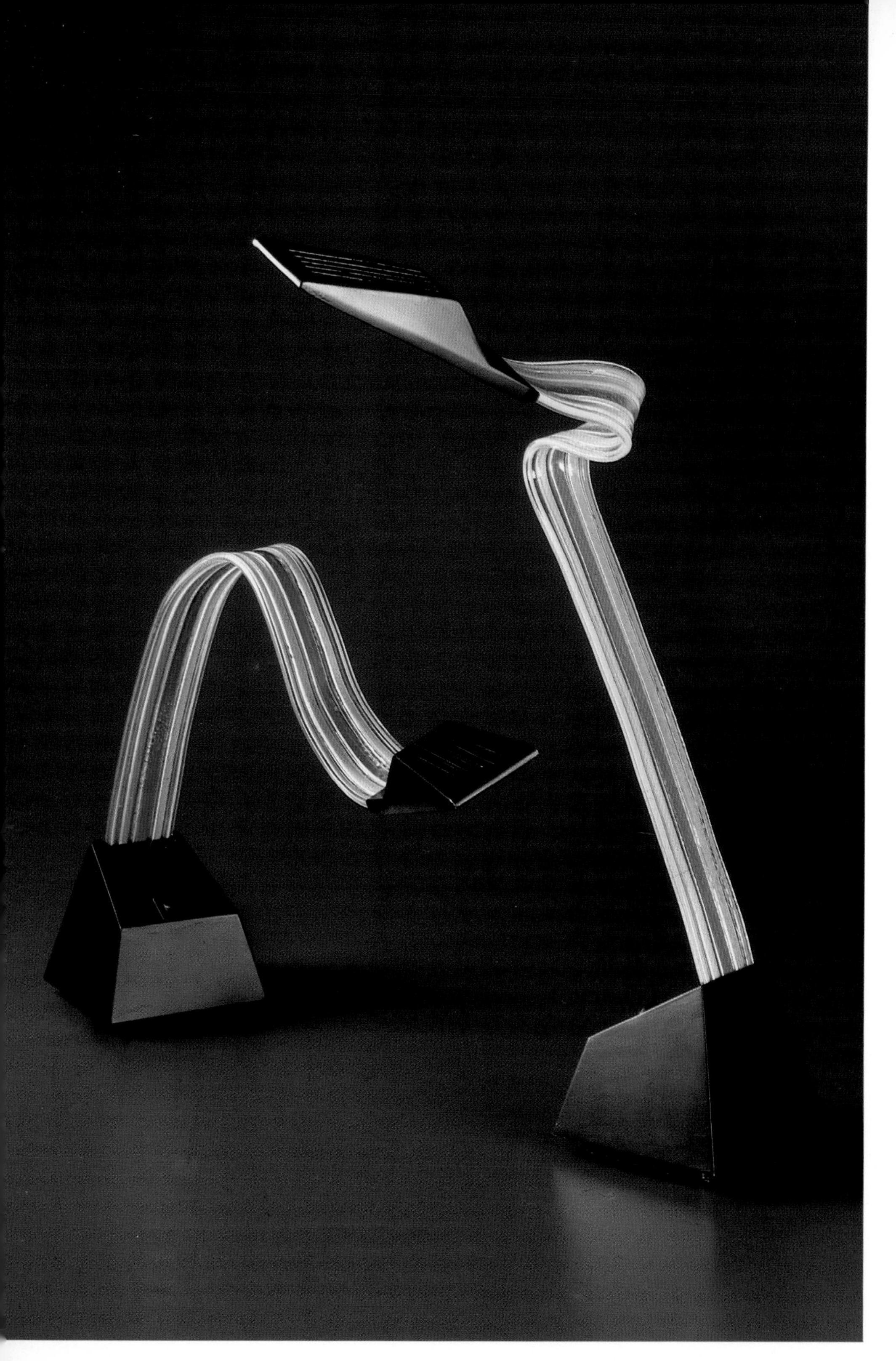

Nastro (Ribbon) halogen desk lamp
Polycarbonate head and base, structural elements are embedded in extruded PVC
Designed by Alberto Fraser
Manufactured by Stilnovo
Italy, 1983

Toy Light Cycle with rip cord, as featured in the film *Tron* by Walt Disney Productions
ABS and synthetic elastomer
Manufactured by Tomy
Japan, 1982

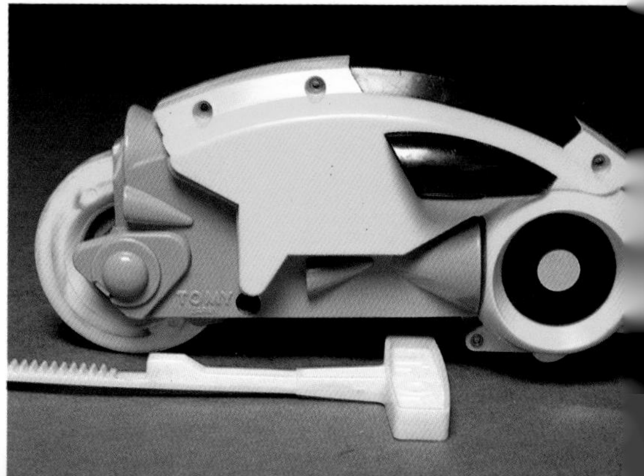

The Nastro light is an unusual concept, spoilt by the ungainly black box housing the transformer. The Light Cycle is a beautiful piece of plastics engineering with a special high-speed mechanism. The rider is inside. It is based on a film visualization by Syd Mead: a cycle made of light. The picnic set fits into a box carried by a high density Polythene strap – it is much lighter and more ingenious than Bandalasta (p. 41).

51-piece picnic set
SAN (styrene acrylonitrile)
Manufactured by Riviera Foods
USA, 1981

Soft-edged hardware: tough, warm, user-friendly ABS can be textured to resist finger-marks and is the perfect material for a switchboard that has its own voice-synthesizer for the blind operator.

Olteco ICS 4000 telephone switchboard
ABS with textured finish
Designed by George J. Sowden
Manufactured by Olivetti
Italy, 1983

Pen holders
ABS
Designed by Rino Pirovano
Manufactured by Rexite di Rino Boschet
Italy, 1983

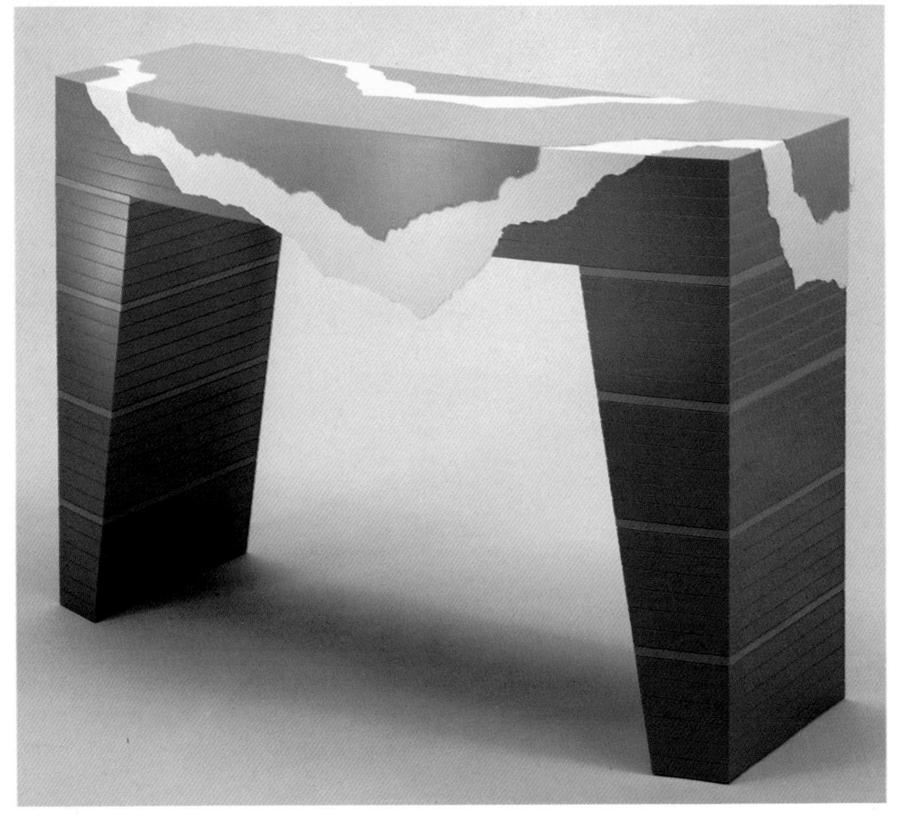

Samson table
Cast polyester resin
Designed by Gaetano Pesce in collaboration with Jean-Luc Muller
Manufactured by Cassina
Italy, 1980

Strata console with draped table cloth
Formica Colorcore melamine
Designed by Brian Faucheux
USA, 1983

Every Samson table is slightly different, according to the catalogue: 'almost rectangular, almost square, almost round'. The angled legs, in particular, deliberately disturb one's sense of equilibrium and structure. At least the table-top is horizontal; its edge appears broken. The tablecloth on the Colorcore console is chipped by hand. The piece looks solid, although it is hollow.

A collector's guide

Advice to collectors

The chief advantage that plastics still enjoy over traditional antiques is that the best places to find them are not necessarily expensive antique shops, but flea markets, jumble sales and relatives' attics and cupboards. Bandalasta teasets and Xylonite hairbrushes are still largely despised, although prices for Art Deco jewellery and Bakelite wireless sets escalated some time ago. On market stalls plastics also have the advantage of surviving the worst weather virtually unharmed. But it has to be said that it is no longer so easy to find them there.

Look first for a product in good condition bearing a mark of some kind, either a patent or registered design number, a trademark or trade name. A moulding with any of these will be of more value as the story of its manufacture can be unearthed at the Patent Office. This is a very exciting part of collecting nineteenth- and twentieth-century industrial archeology, especially if the original drawings or specification can be located.

Look for the best representation possible of the style of the period, combined with an appropriate use of plastics. No matter how 'cheap' the moulding is in terms of apparent value or application, always look for quality in moulding and finish.

The easiest plastics to identify on market stalls are probably celluloid, Bandalasta Ware urea thiourea formaldehyde, Bakelite-type phenolics, acrylic and polythene. But one must progress quickly to learning to identify vulcanite, shellac, horn and casein, always keeping an eye open for objects which appear to be moulded from something strangely organic, which do not feel hard and cold like metal, glass or ceramic, but warm and rather hollow.

Looking after plastics

There is far more ignorance about cleaning and looking after plastics than there is about other materials. The main reason is that most people do not know what their plastics possessions are made of. In addition, plastics are marketed as being virtually self-maintaining, which is, of course, not true. Plastics that are well cared for look better and last longer.

Places of apparently suitable authority provide little guidance on the subject of looking after plastics. The British Plastics Federation, for instance, projects a very fatalistic attitude, stating that if something gets scratched, it is scratched and that is it. On the other hand, The Plastics & Rubber Institute is always helpful and ready to give advice. Not only does it have its own historical collection to look after, but it has launched a joint appeal to build up the London Science Museum exhibit, and can call on members to advise on any historical problem.

Plastics do have the advantage over many other materials that they are much easier to keep clean, and do not rust or tarnish, although hinges and catches may do so. The safest approach is to give a wipe with a soapy cloth, rinse and polish with a dry soft cloth, before trying any other method of cleaning. Never use a gritty, abrasive type of cleaner or wire wool. The surface of many acrylic baths has been ruined by the use of traditional scouring powders, invented originally for metal baths. Instead, use non-abrasive cleaners, which are similar to the rubbing compounds used to rub down GRP mouldings or cellulose paint on cars, and avoid solvent-based cleaners.

Metal polishes are safe to use and produce a very pleasant sheen, but test first in a hidden spot. Fine sandpaper or flour paper can even be used. Aerosol sprays (such as Mr Sheen and Johnson's Wax Sparkle) which incorporate an antistatic agent, are good for polishing phenolic (Bakelite) although they are not recommended for all plastics. Materials such as polythene and polypropylene and tough modern plastics are designed to be scrubbed in hot water; early plastics such as celluloid or casein should never be soaked in water but simply wiped.

Some plastics will alter appearance through the effects of light, heat and ageing. Some fade dramatically, such as ebonite, lighter shades of phenolic (Bakelite) and polythene. Under the cap of an ebonite pen, or the lid of a Bakelite box, the original unfaded colour can often be found.

Celluloid is made from cellulose nitrate which, like its relative gun cotton, evolves nitric acid in a contained space together with camphor vapour. Even some of the earliest pieces of celluloid are still giving off these fumes, which can explode spontaneously. More common is the slow enbrittlement and shrinkage of certain pieces of celluloid. A close look at the school geometry sets made of celluloid in the 1950s (ill. page 79) will show that many of them have shrunk.

Celluloid mouldings should always be stored in a cool, ventilated place. The irony is that clean air-conditioned rooms create static electricity, the bane of any cleaning person: dust particles seem to appear again almost as soon as they have been removed. But better to have healthy, dusty plastics than tacky, disintegrating mouldings.

Fortunately a celluloid explosion is most unlikely, and there is much scaremongering about this by people who take fright at any mention of explosives. However, the nitric fumes might begin to turn the moulding into a sticky mass or powder, and can even rot the wrapping paper quite quickly. If a piece of celluloid has started to rot, try soaking it for two or three weeks in washing soda, or better, urea solution, which should neutralize and stabilize the disintegration. Rinse well and lacquer with clear nail varnish, which is cellulose nitrate or cellulose acetate in solution, allowing several days for drying. The varnish should form a barrier against oxygen. Nail varnish is also useful as a cement for repairing broken celluloid, but with other plastics use an epoxy glue (use sparingly as it tends to go yellow), an instant adhesive, or an adhesive specially formulated for that particular plastic.

At the present time, priceless stocks of early celluloid cinema film are rapidly disintegrating, and in fact half the films made in America before 1950 are now beyond repair. The Motion Picture Academy has set up a film preservation campaign, and it is a race against time to transfer the films on to modern tape. Celluloid as cinema film was more inflammable than when moulded into domestic objects.

Acrylic mouldings such as spectacle frames, boxes and jewellery sparkle beautifully when washed in soapy water (detergent is antistatic), rinsed well and buffed dry. A little metal polish will remove fine scratches as well as polish the surface, and a deep scratch can be removed with various grades of glass paper followed by metal polish. As with phenolic (Bakelite), an aerosol spray polish can be used to give acrylic a shine and to discourage dust. Cellulose acetate mouldings, such as spectacle frames, can be repaired with UHU.

GRP can easily be washed to remove dirt, but if an abrasive cleaner is used, great care must be taken not to rub away the surface gel coat. Most people do not realize that there are special cleaners for removing the stains from melamine cups, the chief complaint about this immensely strong, colourful domestic plastic. Sterilizing solution generally used for babies' bottles is also suitable. Always test first.

Mouldings made of papier mâché can be safely wiped with a soapy cloth, rinsed and dried. A light polish with furniture cream will preserve the surface. Horn can be safely washed in warm water with a little soap, and polished with metal polish or automobile rubbing compound. Tortoiseshell must avoid direct sunlight as the heat will cause it to dry and crack. It can be cleaned with a non-abrasive cream polish or jewellers' rouge.

Simple tests for identification

There is no easy way to identify plastics; even experienced specialists are often mystified by some of the early, unmarked mouldings. As well as physically testing the materials, it is possible by handling different plastics, to develop a feel for their different properties. Many plastics look the same, but after a while small differences can be felt which, together with other clues such as trade marks, smell and function, make possible an intelligent guess at the material and date of the design. A few organizations supply identification kits which contain samples for different polymers, but they do not contain some of the older plastics such as celluloid and shellac. Trips to some of the collections listed later will help a lot.

Below are some simple, non-laboratory tests; they are not conclusive, but are intended as a guide towards identifying materials and mouldings. There are many technical publications on the subject for those more scientifically-minded. If uncertainty and desperation take over, a specialist can be called in.

The moulding must be as clean and pure as possible before carrying out any of the following tests, as grease and dirt can affect the result.

APPEARANCE TEST

Certain plastics are basically transparent (and can also appear translucent or opaque). If an object is transparent, it is most likely one of these:

acrylic (tends to yellow with age)
celluloid (tends to yellow with age)
cellulose esters and ethers (eg cellulose acetate butyrate)
some nylons
polycarbonate
polyesters
polystyrene
PVC (tends to yellow with age)

The following plastics are only translucent or opaque, but never transparent:

ABS (in rare cases, can be clear)
casein
HIPS (high impact or toughened polystyrene)
melamine formaldehyde
certain nylons
phenolics (cast phenolic can be translucent, filled phenolic always opaque and dark in colour)
polypropylene
polythene
polyurethane
natural and hard rubber (ebonite or vulcanite)
shellac
urea formaldehyde and urea thiourea formaldehyde

Marks of the moulding process on the object can indicate whether the plastic is a thermoset or a thermoplastic. For example, a compression moulded box or ash tray will most likely be a thermosetting plastic, and will have a scar of flash line along the part where the two halves of the mould closed together. An injection moulded object, at least if made before the 1970s, will be thermoplastic and will have a sprue mark at the point where the molten polymer was injected into the mould, although the site of the injection could be cleverly disguised. It may also bear the scars of ejector pins, which pushed the moulding out of its mould, and maybe flow or sink marks.

All celluloid objects are made from sheet material of different thicknesses, and many which simulate ivory betray their origin with faint stripes caused by being sliced from laminated blocks. Small solid sculptural forms were stamped or hot pressed from thicker pieces of celluloid.

FEEL TEST

Certain plastics have a characteristic feel about them. Polypropylene and polythene both feel very waxy, but polythene can be scored with a fingernail, unlike polypropylene. HDPE (high density polythene) is not so easy to scratch as LDPE (low density polythene). Polystyrene feels metallic, and when it is tapped it 'rings' with a tinny sound.

FUNCTION TEST

The function of a particular object can give a useful clue to the plastic it is moulded from. The object may come into contact with food, liquid, oil or heat, or perhaps it is simply decorative. Usually the designer has a choice of a few plastics, but some functions will automatically cancel out some of them. There are exceptions, and it is surprising today to come across unlikely black Bakelite (phenolic) kitchen colanders and dark green mottled phenolic tea plates or beakers. The smell of these in use is offensive.

SMELL TEST

Phenolic (eg Bakelite) will usually give off a distinct smell of carbolic acid when wet or warmed. Hard rubber (vulcanite or ebonite) is distinguished from jet and other black materials by its faint yellowish surface bloom, caused by sulphur migration. This can be seen dramatically under the caps and lids of faded pen barrels and boxes. The mouldings also smell of sulphur, even if they are over a hundred years old. If uncertain, put the item inside a plastic bag for a short while and then smell inside the bag, or rub the object against a sleeve.

Sometimes celluloid can also be identified by rubbing the piece briskly against clothing. The frictional heat releases the camphor solvent, which smells very much like moth balls, and this can be sniffed quickly before it disappears.

FLOAT TEST

The density of most plastics does not allow them to float, but polythene and polypropylene are the two main plastics that do so. If polypropylene has been reinforced with a filler, however, it will not float so well. Foamed plastics, such as expanded polystyrene and polyurethane, will obviously float as their cellular structure contains air, and sometimes a moulding may be sandwich construction with a foam layer hidden between solid surfaces.

HEAT TEST (pyrolysis)

This test must be carried out very carefully on a fireproof surface, and fumes must never be inhaled while the plastic is burning, but sniffed quickly afterwards.

Professional laboratories use metal tongs for holding the plastic, but small shavings scraped off in powder form with a very sharp knife from a part of the moulding where it will not be noticeable, can be placed on a saucer or in a glass test tube. The shavings can be ignited by a match or cigarette lighter; in a laboratory the glass test tube is warmed over a bunsen burner or small methylated spirit lamp.

Acrylic: burns with a bluish flame and a floral scent, rather like methylated spirits.

Casein: derived from milk, this plastic burns with an unpleasant but typical smell of burnt milk or cheese.

Cellophane: burns with the smell of burning paper or cotton.

Celluloid: usually the whitish powder 'pops' with minute explositions and burns with a sooty, yellow flame. Catch the smell of camphor before it evaporates; sometimes it is confusingly absent.

Cellulose Acetate: Cellulose acetate sparkles when ignited and gives off the aroma of vinegar (acetic acid). It is made from cellulose treated with acetic acid.

Cellulose Acetate Butyrate: smells of acidic, rancid butter (butyric acid).

Hard Rubber (vulcanite): burns with an orange-yellow flame which is not self-extinguishing, and gives off the nasty, characteristic smell of burning natural rubber.

Nylon: burns with a blue flame, and its aroma has been described as burning 'vegetation, celery or hair'. If touched with a metal point, it can be drawn out into threads.

Phenol formaldehyde (phenolic, Bakelite): when burnt gives off a strong smell of phenol (carbolic acid), as it also does when it is wet.

Polythene and **Polypropylene:** both burn with a smell of paraffin (wax candle), but polypropylene smells more like hot diesel oil.

PVC: burns with black smoke and gives off hydrochloric acid.

PVAC (polyvinyl acetate) and **copolymers of PVC:** smoky and acrid, burn leaving a black ash. PVAC gives off a sweeter acidic smell.

Shellac: burns easily and smells rather like sealing wax, of which it is still a major constituent.

Urea Formaldehyde and **Melamine Formaldehyde:** it is difficult to distinguish between these two plastics. There is no flame when they burn, and both give off the fishy smell of overheated light sockets.

Test laboratories and consultants

France

LABORATOIRE NATIONALE D'ESSAIS
(National Testing Laboratories)
21 rue Pinel
PARIS 13

Netherlands

KUNSTSTOFFEN EN RUBBER
INSTITUUT TNO
(Plastics & Rubber Institute TNO)
PO Box 29
2501–BD THE HAGUE

UK

Dr C A Redfarn & J Bedford
Consulting Chemists
Quality House
Quality Court
Chancery Lane
LONDON WC2A 1HP

Yarsley Technical Centre Ltd
Trowers Way
Redhill
SURREY RH1 2JN

Colleges of polymer technology have testing laboratories, and will usually help with tests. Here are a few in England:

BRUNEL UNIVERSITY
Dept of Non-Metallic Materials
Uxbridge
MIDDLESEX UB8 3PH

CRANFIELD INSTITUTE OF
TECHNOLOGY
Dept of Materials
Cranfield
BEDFORD MK43 0AL

INSTITUTE OF POLYMER
TECHNOLOGY
Loughborough University of Technology
Loughborough
LEICESTERSHIRE LE11 3TU

IMPERIAL COLLEGE OF SCIENCE &
TECHNOLOGY
Exhibition Road
LONDON SW7 2AZ

LONDON SCHOOL OF POLYMER
TECHNOLOGY
The Polytechnic of North London
Holloway Road
LONDON N7 8DB

TROWBRIDGE TECHNICAL COLLEGE
Dept of Science & Engineering
College Road
Trowbridge
WILTSHIRE BA14 0ES

USA

AMERICAN SOCIETY FOR TESTING
MATERIALS (ASTM)
1916 Race Street
PHILADELPHIA
PA 19103

L. J. Broutman and Associates
10 West 35 Street
CHICAGO
IL 60616

Delsen Testing Laboratories Inc
1031 Flower Street
GLENDALE
CA 91201

US Testing Co Inc
1415 Park Avenue
HOBOKEN NJ 07030

Published guides to identification

IDENTIFICATION OF PLASTICS FOR SCHOOLS, Schools Publication No 2, ICI Plastics Division, PO Box 6, Bessemer Road, Welwyn Garden City. HERTS AL7 1HD

IDENTIFICATION OF PLASTICS BY SIMPLE TESTS, Technical Service Note G104, ICI Plastics Division, PO Box 6, Bessemer Road, Welwyn Garden City, HERTS AL7 1HD

OUTLINES OF CLASSWORK FOR TEACHING PLASTICS: THE IDENTIFICATION AND DESCRIPTION OF PROPERTIES, ESPI (Educational Service of the Plastics Institute), Loughborough University, Loughborough, LEICESTERSHIRE LE11 3TU

PROPERTIES OF PLASTICS, the comparison of properties with a chart/guide to non-laboratory tests, Shell Chemicals, Shell Centre, LONDON SE1

SIMPLE MATERIALS TESTING EQUIPMENT, Project Technology Handbook 3, DESIGN WITH PLASTICS, Project Technology Handbook 8, both published by Heinemann Educational/Schools Council and available from ESPI (Educational Service of the Plastics Institute), Loughborough University, Loughborough, LEICESTERSHIRE LE11 3TU

SIMPLE METHODS FOR THE IDENTIFICATION OF PLASTICS by D. Braun, published by Carl Hanser Verlag, Munich 1982; distributed in the USA by Macmillan Publishing Co. Available in England from the Plastics & Rubber Institute, 11 Hobart Place, LONDON SW1W 0HL. A guide with an easy-to-read pull-out Plastics Identification Table by Dr Hans Jürgen Saechtling.

Identification Kits

Plastics Resources Box Samples of raw polymers for experimental work, samples of different manufacturing processes, literature. ESPI (Educational Service of the Plastics Institute), Loughborough University, Loughborough, LEIC LE11 3TU. Price £12.00

The Griffin Polymer Science Kit 14 chemicals (eg acrylic, polystyrene, phenolic) with instruction booklet. Sufficient for 10 experiments with each polymer. A wide range of plastics in small packs is also supplied. Griffin & George Ltd, PO Box 13, Wembley, MIDDX HA10 1LD. Price £47.25.

Plastics Explained Kit of 18 samples of common plastics with 6 booklets for classroom experiments. BP Educational Service, Britannic House, Moore Lane, LONDON EC2Y 9BU. Free.

Resinkit 43 samples of the most common plastics with technical data. Guide to identifying plastics and to recycling plastics. Resinkit, K G Roberts Assoc. Inc. Box 229, East Greenwich, RI 02818, USA. Price $125 in US and Canada, extra postage & handling outside US.

Dealers and stores

ANTIQUARIUS (Antique Market)
135 Kings Road
LONDON SW3

ART DECO
Adalbertstrasse
8 MUNICH 40
WEST GERMANY

AXIS
rue Guénégaud
PARIS 18

BAKELITE
Borgo del Correggio 11
PARMA

CHAINREACTION
251 Camden High Street
LONDON NW1

CICADA
50 Gloucester Road
Brighton
SUSSEX BN1 4AQ

COBRA & BELLAMY
149 Sloane Street
LONDON SW1

COLLEZIONE VITALIA
via Roma 12
SALERNO

STUART CROMBIE
Sutton Coldfield or
Camden Market
Chalk Farm Road
LONDON NW1

FIESTA-GALERIE
103 rue du Cherche-Midi
PARIS 75006

GALERIE 1900-2000
8 rue Bonaparte
PARIS 75006

GRAYS ANTIQUE MARKET
58 Davies Street
LONDON W1

JOHN JESSE & IRINA LASKI
160 Kensington Church Street
LONDON W8

RITA KIM
79 quai Valmy
PARIS 75010

LOFT GALERIE
(ESPACE ROUDILLON)
3 bis rue des Beaux-Arts
PARIS 75006

MANHATTAN AD HOC
842 Lexington Avenue
NEW YORK
NY 10022

MISTINGUETTE
Ickststrasse
8 MUNICH 5

OSIRIS
Türkenstrasse
8 MUNICH 40

PLASTIC WORKS!
1407 Third Avenue
NEW YORK
NY 10024

PRACTICAL STYLING
Centre Point
16/18 St Giles High Street
LONDON WC2H 8LN

ANNA GIANNI RABOLINI
Rodolfo II
via Pontaccio 17
20121 MILAN

SOINTU
20 East 69 Street
NEW YORK
NY 10021

JOE TURNER
Astoria
222 Munster Road
LONDON SW6

TURPIN SANDERS
386 West Broadway
NEW YORK
NY 10012

Glossary

Early materials such as shellac and casein are described in the text.
Words in CAPITALS are in the glossary.

ABS a TERPOLYMER made from three MONOMERS, acrylonitrile, butadiene and styrene. Acrylonitrile and styrene provide chemical resistance (see SAN), butadiene adds impact resistance and makes the plastic suitable for furniture, computer housings etc.

acrylic a hard thermoplastic made from acrylic acid or a derivative of acrylic acid. Best known as a glass substitute under the trade names Perspex, Lucite and Plexiglas.

amino plastics plastics made from ammonia-based compounds, namely UREA FORMALDEHYDE and MELAMINE FORMALDEHYDE.

aniline brown an important dye, originally obtained from indigo, now produced synthetically. It turns brown on exposure to the air, and was used to paint tortoiseshell markings on sheet celluloid.

Bakelite the trade name for plastics produced by Bakelite Ltd in England and Bakelite Corp in America. Still refers to these materials but is frequently used as a generic name for phenol formaldehyde (PHENOLIC).

blow moulding a cheap method for mass-producing hollow objects by blowing air into a hollow tube of softened THERMOPLASTIC POLYMER, so that it expands to line the mould.

casting simple forming of a solid object by pouring resin into a mould, and CURING at room temperature.

Cellophane a Du Pont trade name for film made from regenerated wood pulp (CELLULOSE).

cellular plastics a hard or soft foamed plastic with cells of air caught within the material.

celluloid see CELLULOSE NITRATE.

cellulose the fibrous matter in all plant cells, with a long-chain molecular structure. The most common sources used for making plastics are cotton fibres and wood pulp.

cellulose acetate a tough thermoplastic made from CELLULOSE in the form of cotton linters, treated with acetic acid and acetic anhydride. Widely used as aircraft dope during the First World War, it was welcomed as a fire-resistant alternative to CELLULOSE NITRATE. Now used for many domestic mouldings such as spectacle frames and toothbrush handles, and as transparent packaging film.

Cellulose acetate butyrate a thermoplastic made from CELLULOSE treated with acetic and butyric acids. Transparent, opaque or coloured, with excellent moulding qualitites, it was used where more moisture resistance and dimensional stability than cellulose acetate was required.

cellulose nitrate an inflammable thermoplastic made from CELLULOSE treated with nitric and sulphuric acids. With camphor as PLASTICIZER it was patented as Celluloid in America in 1869.

cold moulded plastics plastics moulded under pressure at room temperature, and cured by heat.

compression moulding the most common method for forming thermosetting resins such as BAKELITE into ash trays, radios etc. The moulding powder, usually with a filler, is placed in the lower 'female' part of a two-part mould. The top part closed on to it, heat and pressure are applied, the plastic flows around the mould and the final moulding is ejected.

copolymer a POLYMER (plastic) made by POLYMERIZING two MONOMERS, eg styrene-acrylonitrile (SAN).

cross-linking the forming of chemical bonds between the molecular chains of a plastic during CURING so that it cannot be re-softened and re-moulded, thus becoming a THERMOSET.

curing the forming of a polymer by POLYMERIZATION and/or CROSS-LINKING.

die steel tool for shaping material by EXTRUSION.

elastomer a synthetic plastic with the flexible properties of rubber.

epoxy (epoxide) resin a very tough THERMOSETTING resin used as a coating, or reinforced to make mouldings or laminates.

ester a compound produced by the reaction between an acid and an alcohol.

extrusion process similar to making spaghetti for moulding plastics into continuous lengths of pipes, rods and profiles. The softened material is forced through a shaped DIE.

filler inert material added to a polymer to improve its properties. Usually in powder or fibre form such as wood pulp, cotton flock and talc.

flash a line of excess plastic forced out of a mould along the PARTING LINE leaving a small ridge. It sometimes has to be filed off.

flow lines patterns visible in a moulding indicating the direction of the flow of the molten plastic in the mould. Usually these patterns are painted over, but they can also be used decoratively.

fuchsin a synthetic dye, also known as magenta, named after the brilliant colours of the fuchsia flower.

GRP glass-reinforced polyester, ie polyester resin strengthened by glass fibres, making the resin, which has no strength of its own, into a very tensile material. Widely used to build boats, furniture and cars.

gun cotton CELLULOSE treated with nitric acid with approximately 13% nitrogen content (as against 11% for CELLULOID). Used as an explosive propellant.

hard rubber the VULCANIZATION process taken to the extreme, when the rubber becomes extremely hard and loses all elasticity.

HIPS high impact polystyrene (see POLYSTYRENE).

injection moulding the most widely used high-speed process for mass producing plastics articles. Granules are heated and forced under pressure into a mould, which can be single cavity for a bucket or chair, or multi-cavity for combs and other small objects.

LLDPE linear low density polyethylene, a new type of low density POLYTHENE.

melamine Melamine formaldehyde, a thermoset produced by

reacting melamine (triaminotriazine) with formaldehyde. A tough, glossy plastic usually strengthened with a filler of wood pulp.

molecular weight the total of the atomic weights of all the atoms forming a molecule. See POLYMER.

monomer a simple, low molecular weight compound. POLYMERIZATION links monomers together to form high molecular weight POLYMERS.

mottle the effect of incomplete blending of coloured moulding powders, originally devised to simulate wood grain, marble and other natural patterns. Powders or granules of different sizes are very effective, and mineral particles and pearl essence create pearly nacreous colours.

nylon not one material but a group of very tough and flexible materials called polyamides. Thermoplastic and usually found as fibres or used solid as gears, zips and more recently, as dyed jewellery.

organic compounds compounds containing carbon in their molecular make-up. Nearly all plastics and rubbers are based on carbon.

parting line a line on a moulding indicating where the halves of a mould closed together.

phenolic shortened version of phenol formaldehyde (see BAKELITE). Phenolic is usually reinforced with a FILLER, but cast phenolic has no filler and can be translucent. It can be easily coloured and was used decoratively for jewellery, radio cabinets and all kinds of ornaments.

plasticizer a substance added to the POLYMER to make it more flexible and easier to mould.

Plastisol PVC paste for coating fabrics, foam furniture and metal parts with a rubbery skin.

polycarbonate a very tough thermoplastic, usually found as a substitute for glass, eg vandal-proof telephone kiosks, bullet-proof shields, baby bottles and picnicware.

polyesters complex ESTER compounds which are thermosetting and can be POLYMERIZED at room temperature, eg GRP.

polymer another word for a plastic material: one which has been made from chains of molecules of one or more MONOMERS. Polymers (plastics) are ORGANIC substances of high MOLECULAR WEIGHT, made from hundreds or thousands of molecules linked together in a repeating chain pattern (also known as macromolecules).

polymerization the chemical process of linking MONOMERS to form new compounds called POLYMERS. For example, ethylene is polymerized into polyethylene (polythene for short).

polypropylene a thermoplastic polymerized from propene, very close to polythene in molecular structure, but harder, stronger and less flexible.

polystyrene a brittle, water-white thermoplastic polymerized from styrene (phenylethylene). The brittleness is overcome by adding some butadiene, which results in toughened polystyrene, also aknown as high impact polystyrene (HIPS), a COPOLYMER of butadiene and styrene. Expanded polystyrene is the rigid white foam used for packaging.

Polythene a Du Pont trade name, used as a generic name in Britain, for polyethylene, the most widely used plastic in the world. Low density polythene is very flexible; high density polythene is more rigid with a higher softening point, and can therefore be sterilized by steam.

polyurethane a rubbery plastic usually found in the form of expanded soft foam for upholstery and rigid foam for insulation, and also as an ELASTOMER. Self-skinned soft foam has a flexible tough skin formed during moulding.

PVAC (PVA) polyvinyl acetate, most familiar as a white wood glue and as emulsion paint.

PVC polyvinyl chloride, a hard, rigid thermoplastic. Plasticized PVC is the softened type, and is mixed with a PLASTICIZER.

PVC-PVAC copolymer vinyl chloride and vinyl acetate copolymerize to form a very tough and flexible vinyl used for inflatable furniture and structures. and pressed into records.

resin a word synonymous with plastic, referring usually to the unPOLYMERIZED material.

SAN styrene-acrylonitrile, much stronger than POLYSTYRENE as it has been modified by the addition of chemical-resistant acrylonitrile.

self-skinned foam soft or rigid foam with a tough surface formed during moulding when foamed cells compact against the inside surface of the mould.

sink mark a depression on the surface of a moulding particularly where there is a rib on the other side. It is caused by contraction of the plastic in the mould.

sprue the part through which plastic flows from the nozzle into the mould during the injection moulding process. The word also means the small, hard piece which remains, and which is then broken off, leaving a characteristic round scar.

terpolymer a COPOLYMER composed of three MONOMERS.

thermoforming the shaping of heat-softened thermoplastic sheet through heat and/or vacuum.

thermoplastic a plastic material which, when softened in a mould under heat and pressure, forms a shape which can be re-softened and re-moulded, eg polythene, acrylic, PVC, nylon.

thermoset a plastic which under heat and pressure polymerizes into a form which cannot be re-softened due to CROSS-LINKING of the molecules. It is therefore used for components such as light fittings, saucepan handles and ash trays.

urea formaldehyde also called urea, a thermosetting AMINO PLASTIC based on the reaction of synthetic urea condensed with formaldehyde.

urea thiourea formaldehyde the earlier and not so tough version of urea formaldehyde.

vacuum forming see THERMOFORMING.

viscose rayon fillaments extruded from a viscous solution of natural CELLULOSE made from wood pulp.

vulcanization the process which makes rubber mouldings elastic and rubbery.

water-white a grade of colour which looks like clear water.

witness marks the scars left on the object by the ejector pins as they push the moulding out of its mould.

Collections *An asterisk indicates a collection open to the public without appointment

BAKELITE MUSEUM SOCIETY Curator Patrick Cook 99 Blackheath Road LONDON SE10	Immense private collection in the process of finding a permanent location
FRITZ BECHT c/o Intomart Noordse Boosje 15 1211 BD HILVERSUM	The leading collection in the Netherlands
*BETHNAL GREEN MUSEUM OF CHILDHOOD Cambridge Heath Road LONDON E2	Good collection of early plastics and rubber dolls
*BOYMANS-VAN BEUNINGEN MUSEUM Mathenesserlaan 18–20 PO Box 2277 3000 CG ROTTERDAM	Small collection of about fifty Bakelite objects
BRITISH INDUSTRIAL PLASTICS LTD PO Box 11 Tat Bank Road Oldbury Warley WEST MIDLANDS B69 4NF	As British Cyanides Ltd, BIP produced Bandalasta. Collection housed in showcase in main office
BROOKES & ADAMS LTD Shady Lane Kingstanding BIRMINGHAM B44 9DX	Small historical display
STUART CROMBIE Sutton Coldfield or Camden Market Chalk Farm Road LONDON NW1	20s, 30s and 50s plastics and kitsch. Also makes own acrylic lamps
HANS & URSULA KÖLSCH Ursulastrasse 9 4300 ESSEN 1	One of the largest private collections in the world, of which two-thirds went on show in Essen, Zürich and Hamburg in 1984
MONTEDISON COLLECTION Curator Anna Rabolini Foro Buonaparte 31 MILAN	Anna Rabolini's historical collection now housed in the Montedison offices
MUSÉE DU PEIGNE ET DES PLASTIQUES 3 rue de la Victoire 01100 OYONNAX	Comprehensive collection of early horn and celluloid combs and machinery. Arbez-Carme collection of celluloid artefacts and experiments
*THE MUSEUM OF LONDON London Wall London EC2Y 5HN	Collection of horn from 17th century to early 20th century, presented by The Horners Company
NATIONAL PLASTICS MUSEUM PO Box 639 LEOMINSTER MA 01453	A national plastics collection has just been started at this address
NETHERLANDS ORGANIZATION FOR APPLIED SCIENTIFIC RESEARCH PO Box 217 2600 AE DELFT	Large collection of plastics from the 1950s
DIE NEUE SAMMLUNG Staatliches Museum für Angewandte Kunst Prinzregentenstrasse 3 8 MUNICH 22	Small collection

N.V. PHILIPS GLOEILAMPENFABRIEKEN PO Box 218 5600–MD EINDHOVEN	Small archive collection of old mouldings
PLASTICS & RUBBER INSTITUTE (PRI) 11 Hobart Place LONDON SW1	Largest collection of Parkesine, given by Parkes' family
RUBBER & PLASTICS RESEARCH ASSOCIATION OF GREAT BRITAIN (RAPRA) Shawbury Shrewsbury SALOP SY4 4NR	Small but interesting collection of historical rubber and processing equipment
*SAC FRÈRES 45 Old Bond Street LONDON W1	Amber merchants always with a fine window display of East European and Russian amber
*THE SCIENCE MUSEUM South Kensington LONDON SW7	Special plastics display on first floor, the best general historic exhibit in England, formerly the collections of John Jesse and Roger Newport. Examples of all early plastics including Parkesine, shellac, bois durci and rubber. Arrangements can be made to see the rest of the collection which is not on display
SMITHSONIAN INSTITUTION 1000 Jefferson Drive SW WASHINGTON DC 20560	Tapes made by the Plastics Pioneers Association are housed there for researchers
Division of Physical Sciences Smithsonian Institution	Collection includes early American celluloid, a Hyatt billiard ball, Baekeland's laboratory papers and samples of Bakelite up to the 1930s. Also early Du Pont nylon
SUFFOLK RECORD OFFICE County Hall St Helen's Street IPSWICH IP4 2JS	Early records of the British Xylonite Co
VESTRY HOUSE MUSEUM Vestry Road Walthamstow LONDON E17 9NH	British Xylonite Co records and catalogues, and the Halex collection of celluloid objects

Patent offices

France

Service de la Propriété Industrielle
26 bis Rue de Léningrad
75800 PARIS

West Germany

Deutsches Patentamt
Zweibrückenstrasse 12
D–8000 MUNICH 2

Italy

Ministero dell'Industria del Commercio
Ufficio Centrale Brevetti
33 Via Vittorio Veneto
00100 ROME

Netherlands

Patent Office
PO Box 5821
2280 HV–RIJSWIJK

UK

Science Reference Library (Holborn Division)
25 Southampton Buildings
LONDON WC2A 1AW

USA

The Patent & Trade Mark Office
WASHINGTON DC 20231

The Patent & Trade Mark Office
2021 Jefferson Davis Highway
ARLINGTON, VA 22201

Libraries

The Picture Library
The Design Council
28 Haymarket
LONDON SW1Y 4SU

Slides from the PLASTICS ANTIQUES EXHIBITION (1977) are available for hire

Imperial College of Science & Technology
Reference Library
Imperial Institute Road
LONDON SW7

Polymer Science Library (Library of the Plastics & Rubber Institute)
London School of Polymer Technology
Holloway Road
LONDON N7

The Royal Institution Library
21 Albemarle Street
LONDON W1

Rubber & Plastics Research Association of Great Britain (RAPRA)
The Library contains the BPF (British Plastics Federation) library
Shawbury
Shrewsbury
SALOP SY4 4NR

Science Museum Library
South Kensington
LONDON SW7 5NH

Science Reference Library (Holborn Division)
25 Southampton Buildings
LONDON WC2A 1AW

Part of the British Library, London. Was formerly the Patent Office Library, opened in 1855, and claims to be possibly the 'largest comprehensive technical reference centre of its kind in Western Europe'. Holds information on about two million British patents, ie every patent published since 1617, and also houses patent details from 38 countries. Patent information can also be obtained at 26 public libraries around the British Isles. The Science Reference Library is in the process of setting up a European Trade Marks Registry.

Centre d'Étude des Matières Plastiques (CEMP)
21 rue Pinel
PARIS 13

Istituto Italiano dei Plastici
16 via Petitti
20100 MILAN

Library of Congress
10 First Street SE
WASHINGTON DC 20540

National Agricultural Library
10301 Baltimore Boulevard
BELTSVILLE MD 20705

National Bureau of Standards
EOI Administration Building
WASHINGTON DC 20234

National Museum of American History Archive
Division of Physical Sciences
Smithsonian Institute
1000 Jefferson Drive SW
WASHINGTON DC 20560

The library and archive contain Baekeland's papers and notebooks.

Plastics organizations

Belgium

Federation of Belgian Chemical Industries
Square Marie Louise 49
B–1040 BRUSSELS

Finland

Muoviteollisuusliitto
Mariankatu 26B
00170 HELSINKI 17

France

Centre de Recherche sur les Macromolécules
6 rue Boussingault
67 STRASBOURG

West Germany

Verein Deutscher Ingenieure
Gesellschaft für Kunststofftechnik
P.O. Box 1139
4000 DUSSELDORF 1

Gesamtverband Kunststoffverarbeitende Industrie (GKV)
Am Hauptbahnhof 12
6000 FRANKFURT AM MAIN 1

Italy

Istituto Italiano dei Plastici (IIP)
16 via Petitti
20100 MILAN

Netherlands

Kunststoffen en Rubber Instituut TNO (KRITNO)
P.O. Box 297
2501·BD THE HAGUE

Nederlandse Vereniging-Federatie Voor Kunststoffen (NVFK)
PO Box 344
3340–AH WOERDEN

Sweden

Sveriges Plastförbund (SPF)
Sveavägen 35–37
S–11452 STOCKHOLM

UK

Institute of Trade Mark Agents
69 Cannon Street
LONDON EC4

ESPI (Education Service of the Plastics Institute)
University of Loughborough
Loughborough
LEICESTERSHIRE LE11 3TU

Plastics Advisory Service (PAS)
5 Belgrave Square
LONDON SW1X 8PH

The Plastics & Rubber Institute (PRI)
11 Hobart Place
LONDON SW1W 0HL

The Rubber & Plastics Research Association of Great Britain (RAPRA)
Shawbury
Shrewsbury
SALOP SY4 4NR

USA

Plastics Education Foundation
c/o Maurice Keroack
PO Box 12443
ALBANY
NY 12212

Plastics Institute of America Inc (PIA)
Stevens Institute of Technology
Castle Point
HOBOKEN
NJ 07030

The Plastics Pioneers Association
PO Box S–1742
c/o The Plastics Institute of America

The Society of the Plastics Industry (SPI)
355 Lexington Avenue
NEW YORK
NY 10017

The Society of Plastics Engineers (SPE)
14 Fairfield Drive
BROOKFIELD CENTER
CT 06805

Journals and magazines

Canada

CANADIAN PLASTICS
Southam Communications Ltd
1450 Don Mills Road
Don Mills
ONTARIO M3B 2X7

PLASTICS BUSINESS
Kerrwil Publications Ltd
443 Mount Pleasant Road
TORONTO M4S 2L8

France

CAOUTCHOUC ET PLASTIQUES and PLASTIQUES ET ENVIRONNEMENTS INFORMATIONS
5 rue Jules-Lefebvre
75009 PARIS

PLASTIQUES FLASH
142 rue d'Aguesseau
92100 BOULOGNE

Italy

INTERPLASTICS
Casa Editrice Tecniche Nuove
via Moscova 46–49
20121 MILAN

MACPLAS
Promaplast srl
Centro Commerciale Milanofiori
1A strada
Palazzo S/2
20090 ASSAGO (MI)

MATERIE PLASTICHE ED ELASTOMERI
Industria Publicazioni Audiovisivi srl
via Verziere 11
20122 MILAN

PLAST
Rivista della Materie Plastiche
Eris spa
Edizione per l'Industria
Piazza della Republica 26
20124 MILAN

West Germany

KUNSTSTOFF BERATER
Umschau Verlag
Stuttgarterstrasse 24
FRANKFURT AM MAIN 1

KUNSTSTOFFE
Carl Hanser Verlag
Kolbergerstrasse 22
PO Box 860420
D-8000 MUNICH 80

KUNSTSTOFFWELT
A M Schoenleitner Verlag
Johann Sebastian Bach Strasse 1
MUNICH 19

UK

BRITISH PLASTICS & RUBBER (incorporating POLYMER AGE)
Maclean Hunter Ltd
76 Oxford Street
LONDON W1N 0HH

EUROPEAN PLASTICS NEWS (incorporating BRITISH PLASTICS)
IPC Industrial Press Ltd
Quadrant House
The Quadrant
Sutton
SURREY SM2 5AS

PLASTICS & RUBBER INTERNATIONAL
Plastics & Rubber Institute
11 Hobart Place
LONDON SW1W 0HL

PLASTICS & RUBBER WEEKLY
PO Box 109
Maclaren Publishers Ltd
Maclaren House
Scarbrook Road
CROYDON CR9 1QH

USA

CONCISE GUIDE TO PLASTICS
R E Krieger Co
PO Box 542
HUNTINGTON
NY 11743

EUROPEAN PLASTICS NEWS
IPC Business Publications
205 East 42 Street
NEW YORK
NY 10017

MODERN PLASTICS and MODERN PLASTICS ENCYCLOPAEDIA
McGraw-Hill Inc
1221 Avenue of the Americas
NEW YORK
NY 10020

PLASTICS
Western Plastics News Inc
1704 Colorado Avenue
SANTA MONICA
CA 90404

PLASTICS DESIGN FORUM
Industry Media Inc
1129 East 17 Avenue
DENVER
CO 80218

PLASTICS ENGINEERING and THE INTERNATIONAL SPEAKER
Society of the Plastics Industry Inc
355 Lexington Avenue
NEW YORK
NY 10017

PLASTICS FOCUS
Plastics Focus Publishing Co Inc
95 Madison Avenue
NEW YORK
NY 10016

PLASTICS WEEK and PLASTICS WORLD
Cahners Publishing Co
221 Columbus Avenue
BOSTON
MA 02116

PLASTICS WEST
465 California Street
SAN FRANCISCO
CA 94104

Trade names

This list is mainly historical up to the late 1940s. Many company names have since changed.

CASEIN

Aralac	National Dairy Products Corp, USA
Akalit	German casein
Ambloid	Japanese casein
Ambroid	Japanese casein
Ameroid	American Plastics Corp, USA
Beroliet	Dutch casein
Casolith	Dutch casein
Coronation	George Morrell Corp, USA
Dorcasine	O. Murray & Co, UK
Ergolith	McLeod & McLeod Ltd, UK
Erinoid	Erinoid Ltd, UK
Estolit	artificial horn made in Estonia
Gala	George Morrell Corp, USA
Galalith	International Galalith, Germany and France
Galorn	George Morrell Corp, USA
Ikilith	S. R. F. Freed, UK
Karolith	American Plastics Corp, USA
Kasolid	Synthetic Plastics Co, USA
Keronyx	Aberdeen Combworks Co, UK
Kyloid	Kyloid Co, USA
Lactilith	casein sheet produced in Belgium
Lactoid	BX Plastics, UK
Lactoloid	Japanese casein
Lactonite	Casein produced in Estonia
Lactophane	British Cellophane Ltd, UK
Maco	Prolamine Products Inc, USA
Pearlalith	B. Schwanda & Sons, USA
Protoflex	Glyco Products Co, USA
Zoolite	Italian casein

CELLULOSE ACETATE

Acele	E. I. du Pont de Nemours, USA
Acelose	American Cellulose Co, USA
Acetyloid	Japanese cellulose acetate
Amer-glo	Celanese Plastics Corp, USA
Armourbex	BX Plastics, UK
Bakelite Cellulose Acetate	Celanese Celluloid Corp, USA
Bexoid	BX Plastics, UK
Celastoid	Celanese Celluloid Corp, USA
Cellastine	Celanese Celluloid Corp, USA
Cellidor	Farbenfabriken Bayer, West Germany
Cellit	J. M. Steel & Co, UK
Cellomold	F. A. Hughes & Co, UK
Cellon	Dynamit-Nobel AG, West Germany
Cellulate	National Plastic Products Co, USA
Charmour	Celanese Corp, USA
Cinemoid	movie film by British Celanese, UK
Clair de Lune	Celanese Celluloid Corp, USA
Clarifoil	British Celanese, UK
Clearsite	Celluplastic Corp, USA
Dorcasite	Charles Horner Ltd, UK
Doverite	Dover Ltd, UK
Dufay-Chromex	Dufay-Chromex Ltd, UK
Duroid	Duroid Covering Co, UK
Ecarit	German cellulose acetate
Embacoid	May & Baker, UK
Embafilm	May & Baker, UK
Enameloid	cellulose acetate and acrylic products by Gemloid Corp, USA
Erinofort	Erinoid Ltd, UK
Fibestos	Monsanto Chemical Co, USA
Firmoid	Bluemel Bros, UK
Gemlite	Gemloid Corp, USA. Also applied to acrylic and polystyrene
Irilit	Danish cellulose acetate
Isoflex	O. & M. Kleeman, UK
Louvreglas	Doane Products Corp, USA
Lansil	Lansil Ltd, UK
Lumapane	Celanese Corp, USA
Lumarith	Celanese Plastics Corp, USA
Lustrac	Lustrac Plastics Ltd, UK
Macite	Manufacturers Chemical Corp, USA
Manusolite	French cellulose acetate
Marolin	Czecho-Peasant Art Co, USA
Nixonite	Nixon Nitration Works, USA
Novellon	British Celanese, UK
Parklite	Parkwood Corp, USA
Parkwood	Parkwood Corp, USA
Plastacele	E. I. du Pont de Nemours, USA
Plasticoil	Schwab & Frank Inc, USA
Plastiktrim	R. D. Werner Co, USA
Plastoflex	Advance Solvents & Chemical Corp, USA
Plastitube	Schwab & Frank Inc, USA
Plastrim	Michigan Molded Plastics Inc, USA
Rexenite	The Rexenite Co. Inc, USA
Rhodoid	M. & B. Plastics, UK
Rhodoid	Rhône-Poulenc, France
R-V-Lite	Arvey Corp, USA
Safety Samson	Celanese Celluloid Corp, USA
Seracelle	Courtaulds, UK
Sicoid	French cellulose acetate
Sundora	E. I. du Pont de Nemours, USA
Tec	Tennessee Eastman Corp, USA
Tenite	Kodak Ltd, UK
Tenite I	Tennessee Eastman Corp, USA
Utex	H. D. Symons & Co, UK
Utilex	Utilex & Co, UK
Vimlite	Celanese Plastics Corp, USA
Vetroloid	British Celanese, UK
Vitapane	Arvey Corp, USA
Vu-Lite	Monsanto Chemical Co, USA
Vuepak	Monsanto Chemical Co, USA
Wireweld	British Celanese, UK

CELLULOSE ACETATE BUTYRATE (CAB)

Hercose	Hercules Powder Co, USA
Rexenite	The Rexenite Co, USA
Tenite II	Tennessee Eastman Corp, USA

CELLULOSE NITRATE (Celluloid, cellulose nitrate film and leathercloth)

Amer-glo	Celanese Plastics Corp, USA
Amerith	Celanese Plastics Corp, USA
Book Tex	Atlas Powder Co, USA
Bexoid	BX Plastics Ltd, UK
Campholoid	Japanese cellulose nitrate
Cascelloid	Cascelloid Ltd, UK

Cascaphane — Cascelloid Ltd, UK
Celastics — Celastic Corp, USA
Cellulac — British Plastoids Co, UK
Celluloid — Celanese Plastics Corp, USA
Celluvarno — Sillcocks-Miller Co, USA
Dentagiene — Canadian Industries Ltd, Canada
Dia-Nippon film — Japanese motion picture film
Dumold — E.I.du Pont de Nemours, USA
Exonite — Dover Ltd, UK
Fiberlac — Monsanto Chemical Co, USA
Fiberloid — The Fiberloid Corp, USA
Fiberlon — The Fiberloid Corp, USA
Filac — Alfred Harris & Co, GB
Flexseal — Flexrock Co, USA
Gemlike — Gemloid Corp, USA
Halex — Halex Ltd, UK
Hercules Cellulose Nitrate Flake — Hercules Powder Co, USA
Herculoid — Hercules Powder Co, USA
Hycoloid — Celluplastic Corp and Hygenic Tube & Container Co, USA
Invaleur — Celanese Plastics Corp, USA
Ivoride — Daniel Spill Co, UK
Keratol — Atlas Powder Co, USA
Kodafilm — Eastman Kodak Co, USA
Kodaloid — Eastman Kodak Co, USA
Lusteroid — Lusteroid Container Co, USA
Mural Rexine — ICI Ltd, UK
Nitron — Monsanto Chemical Co, USA & UK
Nixonoid — Nixon Nitration Works, USA
Oralite — Oralite Co, UK
Pentex — UK Plastics Ltd, UK
Permanite — Parker Pen Co, USA
Phoenixite — Japanese celluloid
Plastine — Sillcocks-Miller Co, USA. Also black cellulose acetate powder made in France.
Protecto — Celluloid Corp, USA
Proxyl — Lee S. Smith & Son Manufacturing, USA
Pyralin — E.I.du Pont de Nemours, USA, but this tradename now refers to their polyimide resin
Pyra-Shell — Shoeform Co, USA
Radite — Shaeffer Pen Co, USA
Rexine — ICI (Rexine) Ltd, UK
Samson — Carpenter Steel Co, USA
Simco — Sillcocks-Miller Co, USA
Viscoloid — E.I.du Pont de Nemours, USA
Xylonite — David Spill Co, later British Xylonite Co, UK
Zaflex — Atlas Powder Co, USA
Zakaf — Atlas Powder Co, USA
Zapon Leathercloth — The Locomotive Rubber & Waterproofing Co, UK

MELAMINE FORMALDEHYDE

Arborite — Arborite Ltd, UK
Beetle Melamine — British Industrial Plastics, UK
Catalin Melamine — Catalin Corp, USA
Coronet — AB Tilafabriken, Sweden
Formica — Formica Ltd, UK and Formica Insulation Co, USA
Gaydon — British Industrial Plastics, UK
Isomin — Perstorp AB, Sweden
Melantine — Ciba Products Corp, USA
Melaware — Ranton & Co, UK
Melbrite — Montedison, Italy
Melmac — American Cyanamid Co, USA
Melmex — British Industrial Plastics, UK
Melolam — Ciba-Geigy, UK
Melopas — Ciba-Geigy, UK
Melurac — American Cyanamid Co, USA
Mepal — Rosti, Denmark
Micarta — Westinghouse Electrical & Manufacturing Co, USA
Perstorp — Perstorp AB, Sweden
Plaskon Melamine — Allied Chemical Corp, USA
Resimene — Monsanto Chemical Co, USA
Resogil — Resopal, France
Resopalit — Resopal, Germany
Resimine — Monsanto Chemical Co, USA
Warerite — Bakelite Xylonite, UK
Watertown Ware — Watertown Manufacturing Co, USA

NYLON

Antron — E.I.du Pont de Nemours, USA
Beetle Nylon — British Industrial Plastics, UK
Bri-Nylon — Courtaulds, UK
Cantrece — E.I.du Pont de Nemours, USA
Caprolan — Allied Chemical Corp, USA
Durethan — Bayer, Germany
Enkalon — British Enka, UK
Maranyl — ICI, UK
Melopas — (polyamide formaldehyde) Ciba Products Corp, USA
Nylon — E.I.du Pont de Nemours, USA, and ICI, UK
Perlon — Bayer, Germany
Plaskon Nylon — Adell Plastics, USA
Renyl C — Montedison, Italy
Rilsan — Organico, France
Sniaform — SNIA, Italy
Sniamid — SNIA, Italy
Ultramid — BASF, Germany
Vestan — Bayer, Germany
Zytel — E.I.du Pont de Nemours, USA

PHENOL FORMALDEHYDE (Phenolic) Including cast phenolic and laminates

Acrolite — Consolidated Molded Products Corp, USA
Aeroplastic — Aeroplastics Ltd, UK
Amberlite — Resinous Products & Chemical Co, USA
Aqualite — National Vulcanized Fiber Co, USA
Aquapearl — Catalin Corp, USA
Arcolite — Consolidated Molded Products Corp, USA
Azolone — Belgian phenolic
Bakelite — Bakelite Corp, USA and Bakelite Ltd, UK
Baker Cast Resin — Baker Oil Tool Co, USA
Beckacite — Beck, Koller & Co, UK
Bondex — McInerney Plastics Co, USA
Carvacraft — J. Dickinson & Co, UK
Catabond — Catalin Corp, USA

Catalex	Catalin Ltd, UK
Catalin	Catalin Corp, USA and Catalin Ltd, UK
Cegeite	French phenolic
Cellanite	Continental-Diamond Fiber Co, USA
Co-Ro-Lite	Columbian Rope Co, USA
Crystle	Marblette Corp, USA
Dekorit	German phenolic
Delaron	De La Rue Plastics, UK
Dilecto	Continental-Diamond Fiber Co, USA
Duraloy	Detroit Paper Products Corp, USA
Durez	W.R.Grace & Co, UK
Durez	Durez Plastics & Chemicals, and Durite Plastics, USA
Elo	Birkby's Ltd, UK
Epok	British Resin Products, UK
Erinite	Erinoid Ltd, UK
Erinoplast	Erinoid Ltd, UK
Fabrolite	The British Thomson-Houston Co, UK
Faturan	German phenolic
Featalak	Featly Products Ltd, UK
Featalite	Featly Products Ltd, UK
Fiberite	Fiberite Corp, USA
Fluosite	Italian phenolic
Formalin	National Plastic Products Co, USA
Formica	Formica Insulation Co, UK and Formica Insulation Co, USA
Frebol	F.Boehm, UK
Futurit	Hungarian and Czechoslovakian phenolic
Futurol	Czechoslovakian phenolic
Gaydon	John Dickinson & Co, UK
Gemstone	A.Knoedler Co, USA
Glasfloss	Durez Plastics & Chemicals Ltd, USA
Heresite	Heresite and Chemical Co, USA
Indur	Reilly Tar & Chemical Corp, USA
Indurite	Indurite Moulding Powders, UK
Ivoricast	Plastics Research Co, USA
Kellite	Kellogg Switchboard & Supply Co, USA
Lamicoid	Mica Insulator Co, USA
Leukorit	German phenolic
Lorival	United Ebonite & Lorival Ltd, UK
Luxene	Bakelite Corp, USA
Makelot	Makelot Corp, USA
Marblette	Marblette Corp, USA
MIC	Molded Insulation Co, USA
Micarta Westinghouse	Westinghouse Electrical & Manufacturing Co, USA
Micoid	Mica Insulator Co, and Watertown Manufacturing Co, USA
Mir-Con	Detroit Paper Products Corp, USA
Monolite	Monowatt Electric Corp, USA
Mouldrite	ICI, UK
Neillite	Watertown Manufacturing Co, USA
Neill Ware	Watertown Manufacturing Co, USA
Nestorite	James Ferguson & Sons, UK
Nobellon	Nobell Plastics Co, USA
Olasal	French phenolic
Opalite	Catalin Corp, USA
Opalon	Monsanto Chemical Co, USA
Panadura	Paramet Chemical Corp, USA
Panelyte	St Regis Paper Co, USA
Panilax	Micanite & Insulators Co, UK
Paxolin	Micanite & Insulators Co, UK
Peton	United Insulator Co,
Phenac	American Cyanamid Co, USA
Phenoid	Mica Manufacturing Co, UK
Phenopreg	Detroit Wax Paper Co, USA
Phenrok	Detroit Wax Paper Co, USA
Philite	Philips Lamps, UK
Plitex	Hood Rubber Co, USA
Pregwood	Formica Insulation Co, USA
Progilite	French phenolic
Prystal	Catalin Corp, USA
Rauzene	US Industrial Alcohol Co, USA
Redmanol	Union Carbide Corp, USA
Reflite	Italian phenolic
Resinox	Monsanto Chemical Co, USA
Resocel	Micafil, UK
Resoform	Micafil, UK
Resophene	French phenolic
Revolite	Revolite Corp, USA
Resinox	Monsanto Chemical Co, USA
Rezinwood	I.F.Laucks Inc, USA
Richware	Makalot Corp, USA
Rivtex	Hood Rubber Co, USA
Rockite	F.A.Hughes & Co, UK
Ryercite	Jos.T.Ryerson & Son, USA
Safetyware	Bryant Electric Co, USA
Sigelit	Czechoslovakian phenolic
Sigit	Czechoslovakian phenolic
Silesit	Polish phenolic
Silesitol	Polish phenolic
Spaulding	Spaulding Fiber Co, USA
Spauldite	Spaulding Fiber Co, USA
Sternite	Sterling Moulding Materials, UK
Super-Beckacite	Reichold Chemicals Inc, USA
Synthane	Synthane Corp, USA
Textolite	General Electric Co, USA
Thermazote	Expanded Rubber Co, UK
Tufnol	Tufnol Ltd, UK
Ucinite	Ucinite Co, USA
Uniplast	Universal Plastics Co, USA
Warerite	Warerite Ltd, UK

POLYESTER RESINS

Bakelite Polyester	Bakelite Corp, USA and Bakelite Ltd UK
Homalite	Homalite Co, USA
Kriston	B.F.Goodrich Chemical Co, USA
Laminac	American Cyanamid Co, USA
Paracon	Bell Telephone Laboratories, USA
Paraplex	Resinous Products & Chemical Co, USA
Plaskon	Libbey-Owens-Ford Glass Co, USA
Polyite	Minerva Dental Laboratories, UK
Selectron	Pittsburgh Plate Glass Co, USA
Thalid	Monsanto Chemical Co, USA
Vibrin	Naugatuck Chemical Division of US Rubber Co, USA

POLYETHYLENE (Polythene)

Alathon	E.I.du Pont de Nemours, USA
Alkathene	ICI, UK
Baylon	Bayer, Germany
Carlona	Shell Chemicals
Cobex	BX Plastics, UK

Eraclene ANIC, Italy
Evazote BXL Ltd, UK
Fortiflex Celanese Corp, USA
Hostalen Hoechst, Germany
Lupolen BASF, Germany
Marlex Phillips Petroleum, USA
Moplen Montedison, Italy
Plastazote BXL Ltd, UK
Poly-Ethylene E.I.du Pont de Nemours, USA
Polythene E.I.du Pont de Nemours, USA
Rigidex BP Chemicals, UK
Sellotape Sellotape Products, UK
Telcothene Telcon Plastics, UK
Thermazote BXL Ltd, UK
Vestolen Huls, Germany
Visqueen British Visqueen, UK

POLYMETHYL METHACRYLATE (Acrylic)

Acraglas The Acraglas Co, USA
Acryloid Resinous Products & Chemical Co, USA
Altuglas Altulor, France
Crystalex Detroit Dental Manufacturing Co, and Rohm & Haas, USA
Crystalite Rohm & Haas Co, USA
Crystolex Kerr Dental Manufacturing Co, USA
Diakon ICI, UK
Duplacryl Coralite Dental Products, USA
Filcryl Portland Plastics, UK
Hexite Hexco Products Inc, USA
Jewelite Pro-Phy-lac-tic Brush Co, USA
Kallodent ICI, UK
Kallodentine ICI, UK
Kallodoc ICI, UK
Livetone Precision Laboratories, USA
Lopac Monsanto Chemical Co, USA
Lucite E.I.du Pont de Nemours, USA
Lucitone E.I.du Pont de Nemours, USA
Lumacryl The Dental Manufacturing Ltd, UK
Ora-Crylic Henry P.Boos Dental Laboratories, USA
Oroglas Lennig Chemicals, UK
Palatex Rockland Dental Co, USA
Palatone Schwab & Frank Inc, USA
Perspex ICI, UK
Plasticarve Schwab & Frank Inc, USA
Plax Methacrylate Plax Corp, USA
Plexiglas Rohm & Haas, Germany and USA
Plexigum Rohm & Haas Co, USA
Plextol German acrylic resin
Polylite Minerva Dental Laboratories, UK
Portex Portland Plastics, UK
Thermolite Oralite Co, UK
Trulite Trudent Products Inc, USA
Vedril Montedison, Italy
Vermonite Vernon-Benshoff Co, USA
Vernonite Rohm & Haas Co, USA
Vitredil Montedison, Italy

POLYSTYRENE

Amphenol American Phenolic Corp, USA
Bendalite Bend-A-Lite Plastics, USA
Bexfoam BX Plastics, UK
Bextrene Bakelite Xylonite Ltd, UK
BP Polystyrene BP Chemicals
Carinex Shell Chemicals, UK
Carex Monsanto Chemical Co, USA
Cyrene Neville Co, USA
Distrene British Resin Products, UK
Intelin International Telephone & Radio Manufacturing Corp, USA
Loalin Catalin Corp, USA
Loavar Catalin Corp, USA
Lustrex Monsanto Chemical Co, USA
Plax Polystyrene Plax Corp, USA
Plexine Rohm & Haas Co, USA
Polyfibre Dow Chemical Co, USA
Polyflex Plax Corp, USA
Polyweld American Phenolic Corp, USA
Portex Portland Plastics, UK
Resoglaz Advance Solvents & Chemical Co, USA
Rexol French polystyrene
Ronilla J.M.Steel & Co, UK
Styraloy Dow Chemical Co, USA
Styramic Monsanto Chemical Co, USA
Styrex Dow Chemical Co, USA
Styrite Dow Chemical Co, USA
Styrofoam Dow Chemical Co, USA
Styron Dow Chemical Co, USA
Styroplas F.A.Hughes & Co, UK
Transpex 2 ICI, UK
Trolitul early German polystyrene
Waterlite Watertown Manufacturing Co, USA

POLYVINYL ACETATE (PVAC)

Flexiplast Foster-Grant Co, USA
Gelva Shawinigan Chemicals, UK
Lustrex Foster-Grant Co, USA
Solvar Shawinigan Products Corp, USA

POLYVINYL BUTYRAL

Butvar Shawinigan Products Corp, USA
Butacite E.I.du Pont de Nemours, USA
Saflex Monsanto Chemical Co, USA

POLYVINYL CHLORIDE (PVC) and vinyl based tradenames

Bexone BX Plastics, UK
Breon BP Chemicals, UK
Chlorovene F.A.Hughes & Co, UK
Cobex British Industrial Plastics, UK
Corvic ICI, UK
Darvic ICI, UK
Elasti-glass S.Buchsbaum Co, USA
Elvax E.I.du Pont de Nemours, USA
Extrudex British Industrial Plastics, UK
Fablon Commercial Plastics, UK
Flamenol General Electric Co, USA
Flovic ICI, UK
Formvar Shawinigan Products Corp, USA
Frostone President Suspender Co, USA
Geon B.F.Goodrich Chemical Co, USA
Glam Pantasote Co, USA
Kenutuf J.F.Kenure, UK
Korolac B.F.Goodrich Chemical Co, USA
Korogel B.F.Goodrich Chemical Co, USA

Koroseal	B.F.Goodrich Chemical Co, USA
Krene	Nut Carbon Co, USA
Lustrac	Lustrac Plastics Ltd, UK
Lustrex	Foster-Grant Co, USA
Luxene	Luxene Inc, USA
Mipolam	Dynamit AG, Germany
Naugahyde	Uniroyal, USA
Numax	RFD-GQ Ltd, UK
Opalon	Monsanto Chemical Co, USA
Perflex	Visking Corp, USA
Permalon	Pierce Plastics Division of Visking Corp, USA
Plastazote	Expanded Rubber Co, UK
Plasticell	BTR Industries, UK
Pliaglas	Pioneer Suspender Co, USA
Plioflex	Goodyear Tire & Rubber Co, USA
Plioform	Goodyear Tire & Rubber Co, USA
Pliolite	Goodyear Tire & Rubber Co, USA
Portex	Portland Plastics, UK
Re-dolite	L.C.Chase & Co, USA
Resolite	Oralite Co, UK
Rasovin	S.S.White Dental Manufacturing Co, USA
Resproid	Respro Inc, USA
Saflex	Monsanto Chemical Co, USA
Syntholvar	Varflex Co, USA
Telcovin	Telegraph Construction & Maintenance Co, UK
Tolex	Textileather Corp, USA
Turex	Textileather Corp, USA
Tygon T	US Stoneware Co, USA
Ultron	Monsanto Chemical Co, USA
Velbex	British Industrial Plastics, UK
Veloflex	Firestone Tire & Rubber Co, USA
Velon	Firestone Tire & Rubber Co, USA
Versiflex	Carbide & Carbon Chemicals Corp, USA
Vinal	Pittsburgh Plate Glass Co, USA
Vinylite	Bakelite Corp, USA and Bakelite Ltd, UK
Vinylseal	Bakelite Corp, USA
Vinyon	Bakelite Corp, USA
Vitrolac	RCA Records, USA
Vybak	British Industrial Plastics, UK
Welvic	ICI, UK

PLASTICS MADE FROM PROTEIN

Cardolite	cashew nut shell plastic, British Resin Products, UK
Maizite	protein coating, Prolamine Products Inc, USA
Maizolith	cornstalks, Iowa State College, USA
Norepol	vulcanized vegetable oil, US Department of Agriculture, USA
Oil Stop	cashew nut shell plastic, Irvington Varnish & Insulator Co, USA
Plastone A & B	cottonseed hull and inorganic filled phenolic, National Plastics Inc, USA
Roburine	natural resin based plastic from France
Soylon	soybean fiber, Ford Motor Co, USA
Valite	bagasse based plastics, i.e. from sugar cane, beets, grapes and olives, Valentine Sugars Inc. and Valite Corp, USA
Zinlac	synthetic shellac based on corn zein, William Zinsser, USA

REGENERATED CELLULOSE

Cellophane	E.I.du Pont de Nemours, USA, and British Cellophane, UK
Kodachrome	Eastman Kodak Co, USA
Kodapak	Eastman Kodak Co, USA
Nymphwrap	Sylvania Industrial Corp, USA
Phanocel	British Cellophane, UK
Sernshi	Japanese cellulosic film
Survival	Sylvania Industrial Corp, USA
Sylphrap	Sylvania Industrial Corp, USA
Sylvania Cellophane	Sylvania Industrial Corp, USA
Tectophane	British Cellophane, UK
Viscacelle	British Cellophane, UK
Wellite	Welwyn Plastics, UK

SHELLAC

Cellulak	Continental-Diamond Fiber Co, USA
Complac	Poinsettia Inc, USA
Compo-Site	Compo-Site Inc, USA
Duranold	Speciality Insulation Manufacturing Co, USA (same tradename also in phenolic)
Electrose	Insulation Manufacturing Co, USA
Harvite	Siemon Co, USA
Kurilac	E.I.du Pont de Nemours, USA
Lacanite	Consolidated Molded Products Corp, USA
Micabond	Continental-Diamond Fiber Co, USA

SYNTHETIC RUBBER

Chemigum	butadiene copolymer, Goodyear Tire & Rubber Co, USA
Dulin	synthetic rubber, Empire Rubber Co, UK
Ethyl Rubber	Hercules Powder Co, USA
Lotol	US Rubber Co, USA
Lycra	polyurethane elastomer, E.I.du Pont de Nemours, USA
Marbo Film	rubber hydrochloride, Marbon Corp, USA
Marbon	rubber hydrochloride, cyclorubber and other materials, Marbon Corp, USA
Neozote	expanded synthetic rubber, Expanded Rubber Co, UK
Perbunan	Standard Oil Co, USA and J.M.Steel, UK
Pergut	chlorinated rubber, J.M.Steel, UK
Pliofilm	clear rubber film, Goodyear Tire & Rubber Co, USA
Ruba	synthetic rubber, Synthetic Latex Corp, USA
Rubide	rubber hydrochloride, Firestone Tire & Rubber Co, USA
Rubprene	neoprene, Firestone Tire & Rubber Co, USA
Tornesit	chlorinated rubber, Hercules Powder Co, USA
Uskol	synthetic rubber, US Rubber Co, USA

UREA FORMALDEHYDE and THIOUREA (includes decorative laminates as well as products)

Aerolite	Aero-Research Ltd, UK
Aminolac	Etablissements Kuhlmann, UK

Arodure	US Industrial Chemicals, USA
Bakelite Urea	Union Carbide Corp, USA and Bakelite Ltd, UK
Bandalasta	Brookes & Adams, UK
Beatl, Beetle	Beetle Products Co, UK and American Cyanamid, USA
Beckamine	Beck, Koller & Co, UK
Beetleware	Beetle Products Co, UK and American Cyanamid, USA
Cibanold	Ciba Products Corp, USA
Daka-Ware	Harvey Davies Molding Co, USA
Durez Urea	Durez Plastics & Chemicals, USA
Formica	Formica Insulation Co, UK and USA
Gabrite	Italian urea formaldehyde
Helomit	Danish urea formaldehyde
Lamicoid	Mica Insulator Co, USA (also a phenolic tradename)
Lorival E	United Ebonite & Lorival Ltd, UK
Marbloid	Japanese urea formaldehyde
Mouldrite U	ICI, UK
Nestor	James Ferguson & Son, UK
Nestorite	James Ferguson & Son, UK
Plaskon	Allied Chemical Corp, USA
Plastaloid	Smith-Gaines Inc, USA
Plyamine	Reichold Chemicals Inc, USA
Pollopas	Etablissements Kuhlmann, Austria and British Industrial Plastics, UK
Prystaline	French urea formaldehyde
Rauxite	US Industrial Alcohol Co, USA
Resopal	German laminate
Rhonite	Rohm & Haas Co, USA
Skanopal	Perstorp AB, Sweden
Scarab	Beetle Products Co, UK
Sibitle	Italian urea formadehyde
Traffolyte	Thomas De La Rue, UK
U Foam	ICI, UK
Uformite	Resinous Products & Chemical Corp, USA
Urac	American Cyanamid Co, USA
Uralite	French urea formaldehyde
Urex	French urea formaldehyde
Urocristal	French urea formaldehyde
Warerite	Warerite Ltd, UK

VULCANITE (ebonite, hard rubber)

Ace	American Hard Rubber Co, USA
Super-Ace	American Hard Rubber Co, USA
Amcosite	Siemens Bros.& Co, UK
Bulwark	Redfern's Rubber Works, UK
Cohardite	Connecticut Hard Rubber Co, USA
Dexonite	Dexine Ltd, UK
Endurance	American Hard Rubber Co, USA
Gallia-Rubber	French ebonite
Keramot	Siemens Bros. & Co, UK
Level Chuck	American Hard Rubber Co, USA
Luzerne	Luzerne Rubber, USA
Mercury	American Hard Rubber Co, USA
Navy	American Hard Rubber Co, USA
Onazote	Expanded Rubber Co, UK
Permcol	British Hard Rubber Co, UK
Resiston	American Hard Rubber Co, USA
Rub-Erok	Richardson Co, USA
Rub-Tex	Richardson Co, USA

Picture credits

Photographers' names in *italic*. **Numerals** refer to page numbers.

17 Victoria & Albert Museum, London, Crown Copyright; **18** (left) Victoria & Albert Museum, London, Crown Copyright; (right) Private collections/*John Kaine;* **19** Plastics & Rubber Institute (PRI), London, and private collection/*John Kaine*; **20** (top) Science Museum, London; (left and right) Science Museum, London, Crown Copyright; **21** (top) Victoria & Albert Museum, London/*Owen & Ferrier*; (lower) RAPRA; **22** Roger Newport/*John Kaine*; **23** Private collections/*John Kaine*; **24** Victoria & Albert Museum, London, Crown Copyright; **25** Science Museum, London; **26** (top) Telcon Plastics/*Roger Newport*; (right) Roger Newport/*Roger Newport*; **27** Sylvia Katz/*Gordon Ferguson*/Shell Polymers; **28** Sotheby Parke Bernet & Co., London; **29** Victoria & Albert Museum, London, Crown Copyright; **30** Sylvia Katz/*Roger Newport*; **31** Sylvia Katz/Phaidon Press/Image; **32, 33** Sotheby Parke Bernet & Co., London; **34** Sylvia Katz/*John Kaine*; **35** Sylvia Katz/*Olly Ball*/Design Council, London; **36, 37** Sylvia Katz/*John Kaine*; **38** Roger Newport/*John Kaine*; **39** Sylvia Katz/*Ken Kirkwood*/*World of Interiors*; **40** Sylvia Katz/*John Kaine*; **41** Sylvia Katz/*Olly Ball*/Design Council, London; **42** (top) Sylvia Katz/*Olly Ball*/Design Council, London; (lower) Sylvia Katz/*John Kaine*; **43** BIP Ltd; **44, 45** BIP Archive/Design Council, London; **46** BIP & Roger Newport collections/*Roger Newport*; **47** BIP/*Roger Newport*; **48, 49** Sylvia Katz/*John Kaine*; **50** Sylvia Katz/*John Kaine*; **51** Sylvia Katz/*Ken Kirkwood*/World of Interiors; **52** Sylvia Katz/*John Kaine*; **53** Galerie 1900–2000; **54** Veronica Manussis/*John Kaine*; **55** (top) Victoria & Albert Museum, London; (lower) Sylvia Katz/*Olly Ball*/Design Council, London; **56** Galerie Loft/*Christian Gervais*; **57** (top) Sylvia Katz/*Olly Ball*/Design Council, London; (lower left and right) Galerie Loft/*Christian Gervais*; **58** Yardley & Co.; **59** Private collection/*John Kaine*; **60, 61** Galerie 1900–2000; **62** Galerie 1900–2000; **63** Design Council, London; **64** Fiesta Galerie; **65** Sylvia Katz/*John Kaine*; **66** Design Council, London; **67** The Museum of Modern Art, New York/*Soichi Sunami*; **68** Sylvia Katz/*John Kaine*; **69** Fiesta Galerie; **70** The Museum of Modern Art, New York. Gift of the manufacturer; **71** Design Council, London; **72** Tupperware Co.; **73** Sylvia Katz/*Gordon Ferguson*/Shell Polymers; **74, 75** Design Council, London; **76** (top) Sylvia Katz/*Ken Kirkwood*/World of Interiors; (lower left) Sylvia Katz/*John Kaine*; (lower right) Galerie 1900–2000; **77** Design Council, London

78, 79 Design Council, London; **80** Galerie Loft/*Christian Gervais*; **81** (top) Sylvia Katz/*John Kaine*; (lower) Sylvia Katz/*Olly Ball*/Design Council, London; **82** Fotofolio/*Joe Steinmetz Studio;* **83** Design Council, London; **84, 85** Sylvia Katz/*John Kaine*; **86** (top) Braun AG; (lower) Design Council, London; **87** Galerie Loft/*Christian Gervais;* **88** (top and lower left) Design Council, London; (lower right) Sylvia Katz/*Roger Newport;* **89** *The Observer,* London; **90, 91** Design Council, London; **92** (top) Sylvia Katz/*Olly Ball*/Design Council, London; (lower) Fiesta Galerie; **93** Galerie 1900–2000; **94** (top) Yardley & Co.; (lower left and right) Design Council, London; **95** Design Council, London; **96** The Boilerhouse, Victoria & Albert Museum, London/Design Council, London; **97** (top) Galerie 1900–2000; (lower) The Boilerhouse, Victoria & Albert Museum, London; **98, 99** Design Council, London; **100** (top) Veronica Manussis/*John Kaine*; (lower) Sylvia Katz/*John Kaine*; **101, 102** Design Council, London; **103** Design Council, London; **104** Sylvia Katz/*John Kaine*; **105** Veronica Manussis/*John Kaine*; **106** (top) Design Council, London/*Photo Masera*; (lower) Mobilier International; **107** Artemide; **108** *The Observer,* London; **109** Design Council, London; **110** (top left) Herman Miller Collection/*John Cook*; (top right) Design Council, London; (lower) Elco; **111** (top) Design Council, London; (lower) Kartell; **112** Paul Clark; **113** Design Council, London; **114** Gufram; **115** Design Council, London; **116** (top left) George J. Sowden; (top right) Crafts Council; (lower) Design Council, London/*David Cripps*; **117** Sylvia Katz/*John Kaine*; **118** Anthony Hopper; **119** Design Council, London; **120** (top) Design Council, London; (lower) Buch & Deichmann/*Bjarnhof & Hviid;* **121** Design Council, London; **122** (top) Olympus Optical Co. (UK); (lower) Design Council, London/*Ian Dawson*; **123** *Falchi & Salvador*; **124** Sylvia Katz/*John Kaine*; **125** George J. Sowden; **126** (top) *Brian Beaumont-Nesbitt*; (lower) Design Gap; **127** Design Gap; **128** (top and lower left) Design Gap; (lower right) Plastirama; **129** One Off/*Jacobi*; **130** Cassina; **131** Vojnovi; **132** (top) Nathalie du Pasquier; (lower) Centrokappa; **133** (left) George J. Sowden; (right) Nathalie du Pasquier; **134** Images of America Inc.; **135** Centrokappa; **136** (top) Cicada/*John Kaine*; (lower left) Sylvia Katz/*John Kaine*; (lower right) Design Council, London/*Olly Ball*; **137** Richard Davies; **138** Braun AG; **139** (top) Design Council, London/*Olly Ball*; (lower) Bostik; **140** (top) Stilnovo; (lower) Sylvia Katz/*John Kaine*; **141** Sylvia Katz/*John Kaine*; **142** Olivetti; **143** Rexite di Rino Boschet; **144** (top) Cassina; (lower) Formica Ltd/*Hedrich Blessing*